我开动了！
栗原晴美的美味料理笔记

［日］栗原晴美 著
牛丹迪 译

华中科技大学出版社
http://www.hustp.com
中国·武汉

有书至美
BOOK & BEAUTY

前 言

既不擅长英语又未曾想要好好学英语的我，之前有幸参演NHK（日本放送协会）面向海外的烹饪节目《您的日本厨房》(Your Japanese Kitchen)。虽然我说得不太好，但好歹也是用英语将自制菜肴展示给了各位观众。那是2007年4月的事情。从那时起，我开始用英语记录自己的食谱，从一点一滴慢慢积累，努力地学习英语，希望有朝一日能够讲出一口流利的英语。

2014年开始，我每年都会去两次美国夏威夷州立大学卡比奥拉尼社区学院教授厨艺。大家通常觉得日式料理工序繁杂且花费时间，为了多少能改变一点大家对日料的这种印象，我希望能够将做饭的快乐传达给大家。

回顾自身经验，其实我觉得英语和烹饪非常相似。学英语的时候，我最先学习的是自己最想表达的短语，然后尝试把它应用在生活中。其实学做饭也是一样，先学会制作一道自己喜欢的菜肴的话，就能感受到做饭的快乐。亲自下厨，既可以为亲朋好友提供一桌子美食，也可以好好招待海外的客人。通过烹饪，我们的交际范围会变得越来越广。

除了在《您的日本厨房》节目中教过的食谱和在夏威夷授课时广受好评的食谱之外，本书中还会介绍一些我个人非常希望外国读者能够学会的精品菜肴。为了方便刚刚接触日式料理的朋友理解，本书会从基础讲起。另外，为了便于海外读者制作，我在食材选择上也下了一番功夫。

这本书也同样适用日本读者。除料理食谱之外，我还会介绍其他内容，比如我生活中不可或缺的小情调、日式烹调器具等。如果这本书能够使日式家常菜肴走上更多家庭的餐桌，我会感到非常高兴。

2018年5月　栗原晴美

Foreword

I had never been good at English, and I did not imagine I would ever study English during my lifetime. However, as it happened, in April 2007 I appeared on "Your Japanese Kitchen", an NHK program broadcast overseas, where I began to introduce Japanese home cooking with my poor English. Since then, I have continued to study English with the goal of being able to write my own recipe and convey myself in English.

I started teaching cooking twice a year at Kapi'olani Community College of the University of Hawaii from 2014. People often say, "Japanese cooking is difficult and has so many procedures", or "Japanese cooking is very time-consuming". I want to change that image and convey that it is fun to cook.

Looking back on my experiences, I have found that English and cooking are, in a sense, very similar. With English, I used to memorize the phrases I like to use the most and began by using them. The same thing can be applied to cooking. Begin by making what you particularly like. You will then gradually become interested in cooking. You will cook for your family and friends or entertain your guests from overseas with your homemade dishes. This way, you can enhance your bonds with other people through cooking.

The recipes in this book include recipes I have introduced to people outside Japan through the program "Your Japanese Kitchen", and the recipes that were popular when I taught in Hawaii. I carefully handpicked the recipes that I especially want people overseas to learn. I have tried to explain Japanese dishes simply, beginning with the basic steps, so that people who have never tried Japanese food will nevertheless be able to understand. I have also tried to choose ingredients that are available overseas.

I'm sure that Japanese people will also find these recipes delightful.In addition to Japanese dishes, I have introduced Japanese cooking utensils, and have shared my thoughts on things I cherish in my daily life. I would be happy if this book could give more people an opportunity to try Japanese home cooking.

Harumi Kurihara, May 2018

前言　Foreword　002

本书使用说明　How to use this book　010

随笔　ESSAY　白色围裙　White apron　008

　　　　　　筷架　*Hashioki* (chopstick rest)　029

　　　　　　樱花　Sakura　093

　　　　　　便当　Bento (box lunch)　152

　　　　　　研磨钵　Mortar　161

　　　　　　漆器　Urushi　212

　　　　　　柚子　Yuzu　253

本书中使用的主要调味品　Basic ingredients used in this book　096

日式烹调器具　Japanese cooking utensils　156

蔬菜的切法　How to cut vegetables　280

序章 | PREFACE |

让我们一起做出美味的米饭吧　Let's cook some delicious rice　012

煮饭方法　How to prepare rice　014

一起捏饭团吧　The rice is cooked-now let's make "Onigiri" (rice balls)　017

饭团的制作方法　"Onigiri" (rice balls)　018

一起制作正宗的日式高汤吧　Prepare dashi in the proper way　020

日式高汤的制作方法　How to make dashi (soup stock)　022

用日式高汤制作味噌汤　Miso soup with dashi　024

豆腐裙带菜味噌汤　Miso soup with tofu and wakame　026

第1章
MEAT & FISH 肉类、鱼类料理

猪肉生姜烧　Ginger pork　032

土豆炖牛肉　"Nikujaga" (beef and potato stew)　037

味噌炖茄子牛肉　Eggplant and beef simmered in miso　041

味噌西冷牛排　Sirloin steak marinated in miso　044

筑前煮　Simmered vegetables with chicken, *Chikuzen*-style　048

奶汁焗烤通心粉　Macaroni gratin　052

葱香炸鸡　Fried chicken with leek sauce　056

和风麻婆豆腐　Japanese-style *mabo-dofu*　060

牛肉可乐饼　Beef and potato croquettes　064

炸猪排　"Tonkatsu" (pork cutlets)　068

味噌青花鱼　Mackerel simmered in miso　072

香煮银鳕鱼　Aromatic stewed sablefish　077

鲑鱼鲜虾丸　Salmon and shrimp *tsukune* meatballs　080

日式腌渍鲑鱼　Deep-fried salmon marinated in *nanbanzu*　084

炸虾排　*Ebikatsu* (shrimp cutlets)　088

2

第2章
VEGETABLES 蔬菜料理

胡萝卜金枪鱼沙拉　Carrot and tuna salad　101

土豆沙拉　Potato salad　104

果仁菠菜　Spinach with peanut sauce　109

焯拌番茄　Tomato *ohitashi*　112

卷心菜沙拉　Coleslaw　116

芜菁猕猴桃薄片沙拉　Turnip and kiwifruit carpaccio　121

脆爽拍黄瓜　Easy pickled cucumber　124

美味豆腐　"Gochiso-dofu" (decorated tofu)　128

千层豆腐　Tofu lasagna　132

四季豆炒肉　Stir-fried string beans and ground pork　136

甜咸土豆　"Kofuki-imo" (salty-sweet flavored potatoes)　141

日式葱香土豆烤菜　Japanese leek and potato gratin　145

炸浸蔬菜　Deep-fried vegetables in *mentsuyu*　148

3

第3章
RICE, NOODLES & MORE 主食

日式牛肉饭　*Gyudon* (beef on rice)　164

亲子盖饭　Chicken and egg on rice　168

生姜饭　Ginger rice　172

猪肉蔬菜什锦饭　Steamed rice with pork and vegetables　176

三色盖饭　Three-color rice bowl　180

炸猪排盖饭　"Katsu-don" (pork cutlet on rice)　184

微波炉红豆饭　Microwaved "sekihan" (azuki beans and rice)　188

散寿司　"Chirashi-zushi"　193

反卷寿司　"Uramaki-zushi" (inside-out sushi rolls)　200

牛肉咖喱　Beef curry　205

茄子干咖喱　Curried rice with eggplant　208

日式甜蛋卷　Sweet dashi rolled omelet　216

什锦拌面　Noodles with shrimp and vegetables　220

笼屉荞麦面　"Zaru soba" (cold soba noodles)　224

煎饺　Gyoza (Chinese dumplings)　229

日式什锦煎饼　Okonomiyaki (savory Japanese pancakes)　233

金枪鱼面包小点　Tuna crostini　237

免揉面包　No-knead bread　240

猪肉蔬菜味噌汤　Pork and vegetable miso soup　245

暖身萝卜锅　Daikon noodle pot　249

第4章
DESSERTS 甜品

调味戚风蛋糕　Chiffon cake　257

鲜奶冻　Panna cotta　260

小仓冰激凌　Ogura ice cream　264

妈妈甜甜圈　Mom's doughnuts　268

松软松饼　Fluffy pancakes　272

零失败芝士蛋糕　Fail-proof cheese cake　276

白色围裙
White apron

　　白色围裙是我重要的工作服。人们往往觉得白色易脏，所以很少使用白色的围裙，但对我来说，这正是它的优点。深色的围裙不会突显污渍，人们很难发现其实它已经脏了，而白色的围裙一旦脏了就很容易被注意到。每次围裙上出现污渍，我都会认真地清洗并且将它熨平。重复这样的过程，其实也是时刻提醒自己以后也要好好珍惜它。

　　这条围裙上还有我名字的刺绣。每次穿上围裙，这个刺绣都会激发我对工作的责任感，以及对美好的新的一天的期待。

A white apron is my important work clothes. People tend to avoid wearing white because it can be a challenge to keep it looking white and stain-free. But this is precisely why it is better. It is harder to notice the stain with a darker colored apron. But with the white one, you will wash and iron it promptly if it stains, and this will motivate you to use it with extra care.
The apron is also embroidered with my name. Therefore, every time I put on the apron, I see my name, and it reminds me that I need to be responsible for my own work. And at the same time, it makes me look forward to another joyful day.

KURIHARA

本书使用说明

- 本书使用的量杯容量为200毫升，
  量匙1大匙=15毫升，1小匙=5毫升。
  以平匙为计量标准，即量匙边缘与所量材料在高度上持平。

- 材料用量（4人份、2人份等）仅供参考。

- 日式高汤默认为昆布、鲣鱼花高汤，做法参阅第022页。

- 酱油默认为浓口酱油，砂糖默认为上白糖，面粉默认为低筋面粉。

- 本书中使用的微波炉的工作功率为600瓦，请根据微波炉的实际功率调整加热时间（＊）。

- 摄氏度和华氏度均为用来计量温度的单位。
  本书中使用的温度单位为日本常用的摄氏度。

- 重量单位用克表示。

- 长度单位用毫米、厘米表示。

- 英文用料表中用tablespoon的缩写tbsp来表示大匙，
  用teaspoon的缩写tsp来表示小匙。
  例）酱油2大匙　2 tbsp soy sauce
  砂糖1小匙　1 tsp sugar

＊若使用700瓦的微波炉，请按0.8倍的时间执行；若使用500瓦的微波炉，请按1.2倍的时间执行。
＊微波炉不可使用金属类容器、含有金属部件的容器、非耐热玻璃容器、漆器、木制品、竹制品、纸制品、耐热温度不满140摄氏度的树脂容器等器皿。使用上述器皿可能会引发故障和事故，敬请注意。
＊使用各种厨具前，请务必熟读产品使用说明书，确保正确操作。

How to use this book

- Cup measurements in this book are 200ml. 1 tablespoon is 15ml, and a teaspoon is 5ml. 1cc is the same as 1ml. Level measures are used.
- Quantities described as "Serves 4" or "Serves 2" should be treated as estimations.
- As described on page 023, dashi is extracted from kelp and dried bonito.
- In this book, soy sauce refers to regular soy sauce (koikuchi); the sugar used is white sugar; flour refers to plain flour.
- Microwave oven times are estimated using a 600W model as reference.
- Fahrenheit (°F) and Centigrade (°C) are both units of temperature measurement, but in this book only the standard Japanese measurement of Centigrade is listed.
- Weight measurements are given in grams.
- Length measurements are given in mm and cm.
- Materials are abbreviated as follows:
  tablespoon = tbsp, teaspoon = tsp.
  Example: "2 tablespoons soy sauce" is indicated as "2 tbsp soy sauce" and "1 teaspoon sugar" is written as "1 tsp sugar".

序章 | PREFACE

让我们一起做出美味的米饭吧
Let's cook some delicious rice

我特别喜欢吃米饭，所以一直在研究如何把米饭煮得更好吃。日本普遍使用电饭煲煮饭，但其实用锅煮饭也简单方便。淘米、煮饭的方法多种多样，在这里我就简单介绍一下我平时的做法。大家也可以试着调整水和米的比例，找到适合自己口味的煮饭方法。

I like Japanese-style cooked rice, which is why I have been making all kinds of efforts to cook delicious rice. Japanese people usually use a rice cooker to cook their rice. Even without this convenient tool, however, you can easily cook rice in a pan or a pot.
There are various ways of preparing rice, but I will introduce my own style in this book. I hope you will find your own favorite style by trying several different methods, adjusting the amount of water used, and so on.

煮饭方法

---

1. 碗中加入2杯大米,加凉水,轻搅均匀,倒掉浑浊的淘米水。
2. 用手掌掌根揉搓米粒,使米粒之间产生摩擦。以流水冲洗,沥干水分。重复此步骤,直至淘米水变透明为止。
3. 将洗净的米放在滤筛上,静置约15分钟。
4. 锅中放米,按1:1的比例加水。可根据个人喜好小幅调整水量,喜好口感偏软的米饭则增加水量,喜好口感偏硬的米饭则减少水量。
5. 盖上锅盖,开火。煮沸后转小火,焖制10~12分钟。
6. 关火,静置约10分钟后打开锅盖,翻搅均匀。

How to prepare rice

1. Put 2 cups of rice in a bowl. Fill with cold water and stir the rice gently with your hand. Discard the cloudy water.
2. Rub the grains gently against each other with the heel of your hand. Rinse under cold running water and drain. Continue rubbing and rinsing until the water becomes clear.
3. Drain the rice in a strainer. Let it stand for 15 minutes.
4. Put the rice in a pan. Add the same amount of water to the pan as rice. For a softer texture, add a little more water; for more texture, use a little less water.
5. Cover the pan and turn the heat on high. Bring to a boil and simmer for 10-12 minutes over low heat.
6. Turn off the heat and let stand for 10 minutes. Remove the lid and stir the rice.

一起捏饭团吧

The rice is cooked—now let's make "Onigiri" (rice balls)

用刚出锅的米饭捏饭团特别好吃。我最喜欢的馅料是海带丝佃煮①。除此之外,饭团馅料多种多样,例如梅干、烤鲑鱼、鳕鱼子、缩缅山椒②等。大家也可以尝试将自己喜欢的食材放入饭团。

Rice balls made of freshly steamed rice taste especially good. My favorite filling is julienned kelp *tsukudani* (boiled down in sweetened soy sauce). There are many different types of fillings to choose from: *umeboshi* (pickled Japanese plum), grilled salmon, cod roe, dried young sardines seasons with *sansho* pepper, and so on. Please feel free to make rice balls with fillings of your taste.

---

① 佃煮:蔬菜海鲜,以酱油、糖、甜料酒等炖煮鱼、贝、肉、藻等,浓厚熟成的日式料理,因起源自日本佃岛地区得名。
② 缩缅山椒:无鳞名吃,即小鱼干煮山椒。

饭团的制作方法

米饭 适量

海苔 适量

盐

[馅料]

梅干 适量

海带丝佃煮 适量

咸鲑鱼（烤制后撕碎）适量

1. 将米饭放在保鲜膜上，使其中心往下凹。根据个人喜好，放入约1大匙的馅料后，在上面放适量米饭，轻捏，取下保鲜膜。

2. 把手放入水中，在掌心涂抹少许盐。将米饭置于掌心，捏成圆形。

3. 在外面包上一层海苔。

"Onigiri" (rice balls)

cooked rice
nori seaweed
salt

[fillings]
umeboshi (Japanese pickled plum)
kombu *tsukudani*
   (kombu kelp simmered in salty-sweet sauce)
salted salmon (grilled and flaked)

1. Place some rice onto a plastic wrap. Make a small dent in the middle, and put the filling of your choice (about 1 tbsp) there. Place more rice on top and shape into a ball. Remove the plastic wrap.

2. Soak your hands in water. Sprinkle a little salt on your palm. Then put the rice on it and shape into a ball.

3. Wrap the rice balls with nori seaweed.

一起制作正宗的日式高汤吧

Prepare dashi in the proper way

　　从昆布和鲣鱼花中提取的日式高汤①是日式料理基础中的基础。现如今，即使是在外国也能轻易买到日式高汤粉末或调料包。一方面，我因为日式料理的普及而感到高兴；另一方面，我也想让大家了解一下现煮的日式高汤是何等的美味。日式高汤的制作方法比大家以为的要简单得多，只要把握好火候和时机，就一定能制作出美味又正宗的日式高汤。

Dashi (introduced in this book) is a Japanese soup stock made from kelp and dried bonito flakes, and is a base of Japanese cuisine. Recently, dashi stocks which come in powder and tea bags have become available overseas, which has helped to spread the popularity of washoku. Although I welcome this situation, I wish first, that you taste the deliciousness of authentic dashi made in the proper way. As long as you exercise care with the heating and timing, anyone can make a tasty dashi – it's easier than you might expect.

---

①又被称作"出汁"。用鲣鱼花、昆布等食材煮制而成，味道鲜美，是日式料理中不可或缺的元素。

日式高汤的制作方法

水 6 杯（1200 毫升）
昆布（长约 10 厘米）1 片
鲣鱼花 40 克

1. 清洗昆布，注意动作要轻。用厨房纸巾擦干，在清水中浸泡约 30 分钟。
2. 锅中加水，加热，在水即将沸腾的时候，捞出碗中浸泡好的昆布并将其放入锅中，水开后加入鲣鱼花，继续煮一会儿，水再次沸腾后关火。
3. 静置，待鲣鱼花自然沉底后，用滤网等工具过滤汤汁。

How to make dashi (soup stock)

6 cups (1200ml) water
10cm-square kombu kelp
40g dried bonito flakes

1. Rinse the kombu lightly, and wipe thoroughly with a paper towel. Soak it in the water for about 30 minutes.
2. Turn the heat on high. Remove the kombu just before the water comes to a boil. Bring the water to a boil, and add the bonito flakes. When it comes to a boil, turn the heat off.
3. Let stand until the flakes sink to the bottom of the pot. Then strain through a sieve.

用日式高汤
制作味噌汤

Miso soup with dashi

成功煮制日式高汤之后，首先试着用它来做一碗热腾腾的味噌汤吧。

我超级喜欢味噌汤，每天都要喝上一碗。我有时会用没吃完的蔬菜制作味噌汤的配料，也经常会在整理或清点冰箱内存货的时候，思考各种各样的食材搭配。这次我想要向大家介绍的是日本家庭中最为常见的豆腐裙带菜味噌汤。

If a good dashi is ready, first start with miso soup.
I like miso soup so much that I eat it each and every day. As ingredients, I often use leftover vegetables from my refrigerator. I check what's in the refrigerator and then clean it out by taking various combinations of ingredients for use in a miso soup.
In this book, I introduce the most commonly made miso soup in Japanese homes, one with tofu and wakame.

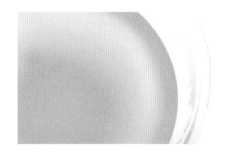

豆腐裙带菜味噌汤

[用料4人份]

绢豆腐①1块
裙带菜（盐渍）30克
日式高汤4杯（800毫升）
味噌4大匙

1. 豆腐切块，每块边长约1厘米。裙带菜泡水，泡除盐分后捞出，切至适当大小，方便食用即可。提前将裙带菜放入碗中。
2. 锅中倒入日式高汤，煮沸后，放入豆腐和味噌酱，搅拌均匀。
3. 把热乎乎的味噌汤倒进装有裙带菜的碗中即可。

---

①绢豆腐：日本将豆腐分为绢豆腐、木棉豆腐等种类。绢豆腐水分充足，质地细腻如绢丝，木棉豆腐在制作过程中会流失一些水分，口感比较紧实。

Miso soup with tofu and wakame

[Serves 4]

1 pack soft tofu
30g wakame seaweed (salt-preserved)
4 cups (800ml) dashi
4 tbsp miso

1. Cut the tofu into 1cm cubes. Soak the wakame in water to take the salt out, and cut it into bite-sized pieces. Place the wakame into a soup bowl.
2. Bring the dashi to a boil. Add the tofu. Add the miso and dissolve.
3. Pour the hot miso soup over the bowl of wakame.

筷架
Hashioki (chopstick rest)

虽说筷架是防止筷尖直接接触桌面的支撑工具，但其实大小不一的筷架也有不同的功能，可以多买一些以备不时之需。我很喜欢尝试用筷架实现各种功能，这让我觉得很有趣。例如，我会用稍微大一点的筷架来盛放柠檬或咸菜之类的食材，有时也会把它们当作小玻璃杯的杯垫来使用。

This tableware is used to place chopsticks on the table. It is handy to have sets of various sizes. I enjoy using them in many different ways. I put garnishes, such as lemon slices or pickles on the larger sized ones. Sometimes I use them as coasters for small glasses.

猪肉生姜烧
Ginger pork

这道菜简单、常见，但这个食谱是我不断尝试、反复摸索的产物。注意一定要提前取出猪肉，让肉恢复常温，并且猪肉蘸取酱汁后要即刻下锅煎制。

This is a simple and easy dish, but I have repeatedly made this dish trying various methods so that I can find an even better taste. The important point is that you take the meat out of the fridge beforehand to have it at room temperature when it is cooked. Also, once you dress the meat with sauce, immediately start grilling it in a pan.

猪肉生姜烧

[用料4人份]

猪上脑肉薄切片 300克
生姜泥 1人匙
酱油 4大匙
甜料酒 3大匙
色拉油 适量
土豆沙拉（参阅第104页）

1. 将酱油、甜料酒、生姜泥倒入碗中混合，放入猪肉片，腌制2～3分钟。
2. 平底锅用大火加热，倒入色拉油，放入猪肉片，煎至猪肉片双面金黄即可。注意在猪肉片表面变色后要立即翻面，动作要快。
   * 如需多次煎制，清洗平底锅后重复此操作即可。
3. 将土豆沙拉和煎好的猪肉片盛入盘中，淋上煎肉汤汁。

* 如果买不到现成的猪肉薄切片，可以趁大块猪肉处于半冷冻状态时自行切片，越薄越好。随后用保鲜膜覆盖肉片，使用擀面杖敲薄。

Ginger pork

[Serves 4]

300g sliced pork shoulder loin
1 tbsp grated ginger
4 tbsp soy sauce
3 tbsp mirin
vegetable oil
potato salad (see page 104)

1. Combine the soy sauce, mirin, and grated ginger. Marinate the pork in the mixture for a few minutes.
2. Heat some vegetable oil in a frying pan over high heat, and briskly sear both sides of the pork slices until brown.
   * If you need to repeat this process several times, make sure to wash the frying pan.
3. Put some potato salad and the ginger pork onto a serving plate. Pour the pan juice from step 2 on the pork.

* If you can't get sliced pork, half-freeze the pork loin and then slice it as thinly as possible. Place each slice between plastic wrap and tap with a rolling pin to make it thinner.

土豆炖牛肉

"Nikujaga"
(beef and potato stew)

　　有一些菜肴，哪怕已经做过无数次，也还是能够在反复制作的过程中发现优化的空间。土豆炖牛肉便是其中之一。看似简单，但要做得好吃并不容易。

　　做好这道菜的秘诀是把土豆炒至表面变透明为止，香味会更加浓郁。

I have some dishes I cook often and yet every time I cook them, I will make a new discovery and I will say, "I have never noticed this important thing before." This is one such dish. It may look easy to cook, but in fact it is difficult to cook it in such a way that it satisfies your taste buds.

The most important thing is to stir-fry the potatoes until their surface becomes translucent. This makes it easier for the potatoes to absorb the flavor of the stock and seasonings.

土豆炖牛肉

[用料4人份]

牛肉薄切片 250 克

土豆 4～5 个（600 克）

洋葱 2 个

日式高汤 1+1/2 杯（300 毫升）

酱油 5～6 大匙

甜料酒 3 大匙

砂糖 4～4+1/2 大匙

酒 1 大匙

色拉油 1 大匙

1. 土豆削皮，切成 4 等份。在水中浸泡五六分钟，捞出，沥干水分。将洋葱切成 4～6 等份，切成梳形块。牛肉片切成一口大小即可。

2. 开中火，平底锅热油。放入土豆，炒至表面变透明。加入洋葱，继续翻炒。

3. 向锅中加入日式高汤、酱油、甜料酒、砂糖、酒，炖煮稍许。撇去浮沫，盖上落盖[①]，炖煮 10～12 分钟至土豆熟软。随后放入牛肉片，注意肉片之间不要相互粘连。肉熟后搅匀，关火。静置入味。

---

①落盖：指小锅盖。日式料理煮炖菜时，经常将比锅盖小一圈的落盖直接盖在食材上，以加速食材入味，缩短烹饪时间，锁住食材的美味。如果没有现成的落盖，可以自行裁剪烘焙纸，制作一次性落盖。

"Nikujaga" (beef and potato stew)

[Serves 4]

250g thinly sliced beef
600g/4-5 potatoes
2 onions
1+1/2 cups(300ml) dashi
5-6 tbsp soy sauce
3 tbsp mirin
4-4+1/2 tbsp sugar
1 tbsp sake
1 tbsp vegetable oil

1. Peel the potatoes and cut each into four pieces. Soak in water for 5-6 minutes and wipe well. Cut the onions into 4-6 wedges. Cut the beef into bite-sized pieces.
2. Heat the oil on medium in a frying pan. Add the potatoes and stir-fry until their surface are translucent. Add the onions and continue to stir-fry.
3. Add the dashi, soy sauce, mirin, sugar, and sake, and simmer. Skim the surface and put a drop-lid on. Simmer for 10-12 minutes until the potatoes are soft. Then add the beef, while spreading each slices, and cook through. Mix well, and turn off the heat and let stand.

味噌炖茄子牛肉

Eggplant and beef simmered in miso

没有一种蔬菜可以像茄子这样好用,煎、炒、烹、炸,样样皆可。

我觉得茄子怎么做都很好吃,接下来介绍的做法是我很喜欢的吃法之一。味噌味道香浓,搭配米饭食用风味更佳。建议茄子切稍大块一些,这样油炸茄子块的口感更佳。

Eggplant is the most convenient vegetable because it can be delicious when cooked in any style, including simmered, grilled, or deep-fried.
I like many eggplant dishes, and this simmered eggplant in savory sauce is one of them. The rich miso taste goes very well with white rice. By cutting the eggplant into larger bite-sized pieces, you can enjoy the texture of the deep-fried eggplant that melts in your mouth.

味 噌 炖 茄 子 牛 肉

[用料4人份]

茄子7～9个（700克）
牛肉片200克

[汤汁]
日式高汤1杯（200毫升）
酱油3大匙
甜料酒3大匙
砂糖3大匙
味噌2～3大匙
豆瓣酱2小匙

煎炸油 适量
色拉油1大匙
生姜泥 适量

1. 制作汤汁，将所需调料混合即可。
2. 将牛肉片切至适当大小，方便食用即可。
3. 茄子去蒂、切成两半，泡入水中去除涩味，捞出擦干。锅中倒入煎炸油，油热后将茄子炸至全熟。
4. 热锅中倒入色拉油，炒牛肉片。
5. 牛肉片炒熟后倒入汤汁，煮开后放入茄子。撇去浮沫，煮炖5～10分钟后关火，静置入味。
6. 连汤汁一起盛入碗中，可根据个人喜好在旁边放适量生姜泥。

Eggplant and beef simmered in miso

[Serves 4]

7-9 eggplants(700g)
200g beef trimmings

[sauce]
1 cup(200ml)dashi
3 tbsp soy sauce
3 tbsp mirin
3 tbsp sugar
2-3 tbsp miso
2 tsp *To-Ban-Jan*

vegetable oil
  for deep-frying
1 tbsp vegetable oil
grated ginger
  --- for garnish

1. Combine the dashi and other ingredients for sauce.
2. Cut the beef into bite-sized pieces, if needed.
3. Cut the stem off the eggplants, cut in half lengthwise, and soak in water to remove the bitterness. Then drain well and pat dry. Heat the deep-frying oil in the frying pan, and deep-fry the eggplants until they are cooked through.
4. Heat the oil in a frying pan and sauté the beef.
5. When the beef is cooked, add the sauce. When it comes to a boil, add the eggplants. Skim the surface and simmer for 5 to 10 minutes. Turn off the heat and let stand.
6. Place in a serving bowl with the sauce. Garnish with some grated ginger to taste.

味噌西冷牛排

Sirloin steak marinated in miso

用味噌腌制食材是日本自古流传下来的传统料理方法。用味噌腌制过的食材不仅味道绝佳，而且易于保存。味噌不仅可以用来腌制肉类，也可以腌制鱼类和蔬菜。为防止牛排煎煳，建议将其切成宽度为2厘米的长条后再进行煎制，这样一来，即使厚度不尽相同，也能够保证每一块肉都受热均匀并煎至最佳状态。也可以根据个人喜好添加绿芥末或七味辣椒粉①。

Marinating ingredients with miso, which is called *misozuke* in Japan, is a traditional way of cooking that we have had in Japan since long ago. It is useful as it not only gives richness to the taste but also keeps the food longer. You can also use fish or vegetables instead of meat.

To avoid the meat getting burnt, cut it into 2cm-wide slices. This way, you can remove each piece from the heat just in time when it is well cooked.If you like, serve with grated wasabi or *shichimi* (seven-flavor chili pepper).

---

① 七味辣椒粉：简称七味，是一种以辣椒为主料的调味料，将辣椒、芝麻、陈皮、芥籽、油菜籽、麻仁、花椒等香辛料研碎后混合而成。

味噌西冷牛排

[用料4人份]

西冷牛排（2厘米厚）4块
绿芥末、酸橘 各适量

[**味噌酱汁**]
味噌 400克
酒 1/2杯（100毫升）
甜料酒 1杯（200毫升）
砂糖 60～80克

1. 制作味噌酱汁。在锅中放入味噌、酒、甜料酒、砂糖，用小火熬20分钟左右，至酱汁黏稠即可。熬制酱汁的过程中注意要不停地搅拌，以防煳锅。
   *这份味噌酱汁可在冰箱冷藏保存近3周。

2. 在牛排的正反两面涂抹上调制好的味噌酱汁，每面的酱汁用量为2大匙。用保鲜膜包好放入冰箱，冷藏1～2天。
   *也可以直接冷冻保存。

3. 取出牛排，用刀或其他工具刮掉牛排表面的味噌酱汁。将牛排切至适当大小，方便食用即可。

4. 将牛排放在烤架上，双面烤制。装盘，可根据个人喜好在旁边摆放喜欢的炒蔬菜、绿芥末、酸橘等。

Sirloin steak marinated in miso

[Serves 4]

4 sirloin steaks
(2cm thick)
wasabi and sudachi
--- to serve

[miso marinade]
400g miso
1/2 cup(100ml) sake
1 cup(200ml) mirin
60-80g sugar

1. Make the miso marinade: Combine the miso, sake, mirin and sugar in a pan and bring to a boil. Keep stirring over low heat to avoid burning, and boil down for about 20 minutes until the sauce thickens.
   * This marinade can keep for up to 3 weeks in the refrigerator.
2. Spread 2 tablespoons of the marinade on each side of the steaks and cover with plastic wrap. Let them stand in the refrigerator for 1-2 days.
   * They can be kept in the freezer as they are.
3. Remove the marinade from the steaks with a spatula and cut into bite-sized pieces.
4. Grill the steaks on both sides. Serve with stir-fried vegetables, wasabi, and sudachi.

筑前煮

Simmered vegetables with chicken, *Chikuzen* style

筑前煮①是日本正月年节菜中经常出现的菜肴，也是我一年四季都经常制作的家常菜。这道菜类似于乱炖，为了保证每一种食材都能够均匀受热，需要将食材切至相近大小，并且按顺序将食材放入锅中。

另外，因为我带有私心，想通过这道菜向大家介绍一些实用的切菜方法（滚刀块切法、银杏叶形切法、半月形切法等），所以特意把每种蔬菜都切成了不同的形状。

This is one of the popular dishes served at New Year's. In my home, however, I cook this throughout the year. The important point in cooking this dish is to cut the ingredients into evenly-sized pieces or slices, and to put them into a pan in the correct order so that they cook evenly.

Because I wanted to introduce to you various Japanese ways of cutting vegetables, I used several techniques.

①筑前煮：因起源自日本九州北部筑前地区而得名，主要原料有鸡肉、胡萝卜、牛蒡、蒟蒻等。

筑 前 煮

[用料 4 人份]

鸡腿肉（无骨）250 克
牛蒡 1 根（180 克）
胡萝卜 1 根（200 克）
煮竹笋 1 小根（150 克）
莲藕 1 根（200 克）
蒟蒻 1 块（200 克）
干香菇 4 个
色拉油 1～2 大匙
日式高汤 1+1/2 杯（300 毫升）
酱油 4 大匙
砂糖 4 大匙
甜料酒 2 大匙
酒 2 大匙

1. 用尽量少的水泡发干香菇，泡发后轻轻地挤去多余水分，去蒂，切成 4 等份。
2. 鸡腿肉切至一口大小。
3. 牛蒡去皮，斜切成 2 厘米厚的块，泡入水中去除涩味后，捞出，沥干水分。
4. 胡萝卜去皮，切成 2 厘米厚的半月形。煮竹笋以滚刀块切法切成块。
5. 莲藕去皮，切至 2 厘米厚的银杏叶形。泡入水中去除涩味后，捞出，沥干水分。
6. 将蒟蒻放入热水中，快速焯水，用漏勺捞起沥干水分。晾凉后切成一口大小。
7. 加热深口平底锅，放入色拉油，炒制鸡肉。依次放入牛蒡、胡萝卜、蒟蒻、香菇、煮竹笋、莲藕，放入每种食材后适当翻炒。如有需要，可根据情况少量加油。
8. 将日式高汤、酱油、砂糖、甜料酒、酒倒入锅中，煮开后撇去浮沫，盖上落盖，煮炖约 15 分钟至汤汁收干。

Simmered vegetables with chicken, *Chikuzen*-style

[Serves 4]

- 250g boneless chicken thighs
- 1 burdock (180g)
- 1 carrot (200g)
- 1 small bamboo shoot (150g), boiled
- 1 lotus root (200g)
- 1 block of konnyaku (200g)
- 4 dried shiitake mushrooms
- 1-2 tbsp vegetable oil
- 1+1/2 cups (300ml) dashi
- 4 tbsp soy sauce
- 4 tbsp sugar
- 2 tbsp mirin
- 2 tbsp sake

1. Soak the dried shiitake mushrooms in just enough water to cover, take the time to let it soften. Squeeze lightly and cut the stems off, then cut into 4 pieces.
2. Cut the chicken into bite-sized pieces.
3. Peel the burdock and cut diagonally into 2cm-thick pieces. Soak in water and drain well.
4. Peel the carrot and cut into 2cm-thick half-moons(*hangetsu-giri*). (see page 280) Cut the bamboo shoot into *ran-giri* pieces. (see page 281)
5. Peel the lotus root and cut into 2cm-thick quarter-rounds(*icho-giri*). (see page 280) Soak in water and drain well.
6. Blanch the konnyaku and drain well. Tear into bite-sized pieces when cool.
7. Heat the oil in a deep frying pan and stir-fry the chicken. Add the burdock, carrot, konnyaku, shiitake mushrooms, bamboo shoot and lotus root and stir-fry in this order. Add a little more oil if necessary.
8. Add the dashi, soy sauce, sugar, mirin, sake. When it comes to a boil, skim the surface and put a drop-lid on. Simmer for about 15 minutes or until the sauce is reduced.

奶汁焗烤通心粉

Macaroni gratin

通心粉、虾、洋葱、蘑菇、鸡肉制成的奶汁烤菜是极为常见的日式西餐,可以说走进了日本的千家万户。在我心中,可以说它是日本人创造的最高杰作。

记得加入足量的奶酪,快要溢出来的那种程度最好,奶酪的量是影响味道的重要因素之一。这道料理和米饭也非常搭,我很喜欢滴上几滴酱油之后和米饭一起吃。

This is a common Western-style dish that many Japanese families prepare at home. This gratin contains macaroni, shrimp, onion, mushrooms, and chicken. I have to say this Japanese-style gratin is one of the best dishes the Japanese have created.
Topping it with plenty of cheese will make this more delicious. It also goes well with steamed white rice. I usually put a little soy sauce on this gratin so it goes even better with white rice.

奶汁焗烤通心粉

[用料4人份]

鲜虾 250克
鸡腿肉 1块
洋葱 1/2个
蘑菇罐头 1罐（100克）
通心粉 100克
奶酪（比萨用）150克

[白酱]

黄油 40克
面粉 50克
牛奶 2+1/2杯（500毫升）
生奶油 1杯（200毫升）
盐 1/2～1小匙
胡椒 少许

色拉油、盐、胡椒 各适量

[准备工作] 烤箱预热至230摄氏度。

1. 洋葱切薄片。
2. 鲜虾剥壳，去虾线。洗净后擦干水分。
3. 鸡腿肉切至一口大小。
4. 取出蘑菇，沥干汤汁。
5. 制作白酱。平底锅中放入黄油，待黄油完全融化后，加入面粉，用小火炒制2～3分钟，注意不要煳锅。少量多次加牛奶，熬制5分钟左右，直至酱汁变黏稠，注意加牛奶的时候要一直搅拌。加生奶油，熬制稍许，加盐、胡椒调味。
6. 煮通心粉，参照包装说明即可。煮好后用漏勺捞出备用。
7. 平底锅中倒入少量色拉油，油热后炒制虾，轻撒盐和胡椒后出锅。可根据情况少量加色拉油，按顺序将鸡肉、洋葱、蘑菇放入锅中并炒匀后，加盐和胡椒。
8. 把步骤7炒好的食材和步骤6煮好的通心粉放入白酱中，拌匀。
9. 盛入耐热容器，撒上奶酪，在预热至230摄氏度的烤箱中烤制15～20分钟。

Macaroni gratin

[Serves 4]

250g shrimp
1 boneless chicken thigh
1/2 onion
1 tin of mushrooms (100g)
100g macaroni
150g pizza cheese

[white sauce]
40g butter
50g flour
2 +1/2 cups (500ml) milk
1 cup (200ml) heavy cream
1/2-1 tsp salt
pepper

vegetable oil, salt,
  and pepper

[preparation] Heat the oven at 230°C.

1. Thinly slice the onion.
2. Remove the shells and devein the shrimp. Wash them well and wipe off the moisture thoroughly.
3. Cut the chicken into bite-sized pieces.
4. Drain the mushrooms.
5. Make the white sauce: Melt the butter in a frying pan, add the flour, and stir for 2-3 minutes. Make sure it doesn't burn. Pour in the milk gradually, and constantly stir for about 5 minutes until it thickens. Add the heavy cream and cook for a short time. Season with salt and pepper.
6. Boil the macaroni according to the time marked on the package. When it's done, drain it in a colander.
7. Heat the oil in a frying pan and stir-fry the shrimp. Season with a little salt and pepper. Empty them into a dish. Add a little more oil if necessary and stir-fry the chicken, onion, and mushrooms in this order, and season them with salt and pepper.
8. Add the chicken, onion, mushrooms, shrimps, and macaroni to the white sauce, and mix together.
9. Place it on a heat-resistant dish, sprinkle the cheese on top, and bake it in an oven at 230°C for 15-20 minutes.

葱香炸鸡

Fried chicken with leek sauce

　　这道菜承载了我和父亲的回忆。我开发这个食谱的初衷是为了让不爱吃鸡肉的父亲也能够喜欢吃。葱香炸鸡也是我众多食谱中当之无愧的人气王。做出酥脆炸鸡的秘诀主要有三点：提前从冰箱中取出鸡肉，使其恢复至室温；严格按食谱用量调制底味；炸制鸡肉前大量蘸取片栗粉。为了防止鸡肉表皮上色但里面没熟的尴尬情况发生，我一般都会炸制两次。

This dish has a lot of memories for me, as I made this for my father who was not fond of chicken. This is one of the most popular dishes among my numerous recipes. The secrets to making crispy fried chicken are to take the chicken out of the fridge and bring it to room temperature before cooking, measure the exact amount of seasonings which are used for the chicken beforehand as instructed, and coat the chicken with plenty of potato starch just before frying it. Fry the chicken twice to avoid the center of the chicken being only half-cooked.

---

①片栗粉：片栗粉原指由猪芽花根茎制成的淀粉，但现在市面上常见的多为土豆淀粉。

葱香炸鸡

[用料 4人份]

鸡腿肉 2块
酱油 1/2大匙
酒 1/2大匙
片栗粉 适量
煎炸油 适量

[葱酱]

葱 1根
色拉油 1/2大匙
红辣椒圈 1个辣椒的量

[调味汁]

酱油 1/2杯（100毫升）
酒 1大匙
醋 2大匙
砂糖 1+1/2大匙

1. 提前从冰箱中取出鸡腿肉，使其自然恢复常温。用叉子在鸡皮上扎一些小孔。鸡腿肉对半切开，用酱油和酒适当腌制。

2. 制作葱酱。先调制调味汁，将酱油、酒、醋、砂糖倒入碗中，搅拌均匀即可。切葱末，先用刀尖在大葱表皮留下密集的划痕，再从一端开始将其切碎。平底锅中倒入色拉油，油热后放入葱末和红辣椒圈快速翻炒，炒匀后倒入调味汁，关火。

3. 煎炸油加热至180摄氏度。鸡腿肉蘸取大量片栗粉，裹匀后下锅炸2～3分钟。用炸网捞出鸡腿肉，静置约4分钟，通过余温使鸡腿肉熟透。随后再次开大火炸制1～2分钟。

4. 鸡腿肉沥干油分后，切至适当大小，方便食用即可。盛盘，淋上葱酱即可食用。

Fried chicken with leek sauce

[Serves 4]

2 chicken thighs
1/2 tbsp soy sauce
1/2 tbsp sake
potato starch for coating
oil for deep-frying

[leek sauce]
1 Japanese leek
1/2 tbsp vegetable oil
1 red chili pepper, chopped
[combined seasonings]
1/2 cup(100ml) soy sauce
1 tbsp sake
2 tbsp vinegar
1+1/2 tbsp sugar

1. Allow the chicken to reach room temperature before cooking. Pierce the chicken skin in several places with a fork, cut the chicken in half, and marinate in the soy sauce and sake.
2. Make the leek sauce: Mix the ingredients for the combined seasonings. Pierce the leek with the tip of a knife all over. Then, chop it finely starting from the end. Heat the oil in a frying pan. Stir-fry the leek and red chili. Stir constantly and add the combined seasonings. Turn off the heat.
3. Heat the oil to 180°C. Cover the chicken completely with potato starch. Deep-fry the chicken for 2-3 minutes. Put it on a rack and leave for about 4 minutes while it continues to cook with the residual heat. Then deep-fry it for 1-2 minutes over high heat again.
4. Drain and cut the chicken into bite-sized pieces. Place on a serving plate and pour the leek sauce over the top.

和风麻婆豆腐

Japanese-style *mabo-dofu*

  我超级喜欢吃麻婆豆腐。这道菜用日式高汤代替中式高汤，口感更为温和。

  我一般会多炒一点肉馅，将炒制好的肉馅放入冰箱冷冻保存，下次想吃的时候可以直接拿出来用，非常方便。

I love *mabo-dofu*. This one uses Japanese-style dashi as a substitute for Chinese soup, giving it a mild flavor.
The meat *an*, before tofu is added, freezes well. So it is convenient to keep in the freezer. All you have to do is add some tofu before eating.

和风麻婆豆腐

[用料4人份]

绢豆腐 2块（约700克）
猪肉、牛肉混合肉馅 200克
蒜末 1大匙
姜末 1大匙
葱末 4大匙

[汤汁]
日式高汤 1+1/2杯（300毫升）
酱油 4～5大匙
甜料酒 2大匙
砂糖 1大匙

盐 少许
片栗粉、水 各1大匙
色拉油 2大匙
芝麻油 适量
红辣椒圈 2～3个辣椒的量
山椒粉 适量

1. 大蒜、生姜切成碎末后，各取1大匙。
2. 葱切成碎末后，取4大匙。
3. 豆腐切成1.5厘米见方的小块。锅中加大量水，烧开，加少许盐后放入豆腐，煮大约2分钟后用漏勺捞出。
4. 将水和片栗粉按照1:1的比例调制成水淀粉。
5. 将汤汁所需调味料倒入锅中，加热备用。
6. 平底锅中放入色拉油，热油，将蒜末、姜末和葱末炒出香味后放入混合肉馅，继续炒制。肉馅炒熟后倒入汤汁，煮开后加水淀粉，增加汤汁的浓稠度。放入豆腐轻轻搅匀后，淋一些芝麻油。
7. 装盘，可根据喜好添加红辣椒圈和山椒粉。

Japanese-style *mabo-dofu*

[Serves 4]

2 packs soft tofu (approx.700g)

200g ground beef and pork mixture

1 tbsp garlic (finely chopped)

1 tbsp ginger (finely chopped)

4 tbsp Japanese leek (finely chopped)

[sauce]

1+1/2cups (300ml) dashi

4-5 tbsp soy sauce

2 tbsp mirin

1 tbsp sugar

salt

1 tbsp potato starch

1 tbsp water

2 tbsp vegetable oil

sesame oil

2-3 red chili peppers(chopped)

sansho powder --- for garnish

1. Finely chop the garlic and ginger.
2. Chop up the Japanese leek.
3. Cut the tofu into 1.5cm cubes. Bring a pot of water to a boil, add a little salt, and boil the tofu in it for about 2 minutes before pouring it into a colander.
4. Dissolve the potato starch in an equal amount of water.
5. Combine the ingredients for the sauce in a pan and heat it.
6. Heat the vegetable oil in a frying pan and stir-fry the garlic, ginger and Japanese leek. Once they become fragrant, stir-fry the ground beef and pork mixture. When the meat is cooked through, pour the sauce in and when it comes to a boil, add the dissolved potato starch and let it thicken. Add the tofu, and mix gently. Sprinkle sesame oil around the edge of the pan.
7. Serve on a plate, garnish with slices of red chili pepper or *sansho* powder if preferred.

牛肉可乐饼

Beef and potato croquettes

这道菜的要点是用黄油炒制洋葱和牛肉后，连汤带料将其倒入土豆泥中。这样可以使洋葱炒肉的香气和土豆泥的口感融为一体。另外，为了进一步丰富口感，建议洋葱稍微切大块一点，并且不要炒得太熟。

土豆是这道菜的主角。到了土豆正当时的季节，或者刚刚购入新土豆的时候，大家一定要趁着食材新鲜尝试制作这种可乐饼。土豆推荐使用男爵土豆①等比较松软的品种。

Cook the onion and beef with butter, and add them to the potatoes along with the gravy so as to transfer their flavor and aroma to the potatoes. The onion should not be cut too finely, nor should it be over-cooked. This will help to ensure that you can still appreciate the texture of the onion.

As potatoes are a main feature of this dish, try this recipe when potatoes are in season or use fresh potatoes from the store. I particularly recommend soft-textured potatoes such as the *Danshaku* variety.

---

①男爵土豆：日本明治时代（1868—1912年）后期，由川田龙吉男爵从美国引进日本的土豆品种。早熟，肉白色，味美。

牛肉可乐饼

[用料 20 个]

土豆 4 个（500 克）
洋葱 1 个（200 克）
牛肉片 200 克
盐、胡椒 各适量
黄油 30 克

[面衣]
面粉、蛋液、面包糠 各适量

煎炸油 适量

[卷心菜沙拉]
卷心菜（切丝）200 克
柠檬汁 1 大匙
橄榄油 2 大匙
盐、胡椒 各少许

[可乐饼蘸汁]
炸猪排酱汁 3 大匙
番茄酱 2 大匙

黄芥末酱 适量

1. 土豆去皮，切成 4~6 等份，泡入清水中以去除部分淀粉，捞出，沥干水分。在耐热容器内铺上厨房纸巾，放入土豆块。用保鲜膜包住容器口，注意不要包得太紧，用微波炉（600 瓦）加热 5~6 分钟至土豆变软。

2. 取下保鲜膜，取出厨房纸巾，趁热捣碎土豆，自然放凉。
   *如不使用微波炉，则需用锅煮熟土豆。

3. 洋葱切碎，牛肉片切碎。锅中放入黄油，加热，黄油融化后放入牛肉，适当翻炒后放入洋葱，继续炒制。撒 1/2 小匙的盐和少许胡椒。

4. 将炒好的牛肉和洋葱连同汤汁一起倒入土豆泥中。用盐和胡椒调味。

5. 将步骤 4 的洋葱牛肉土豆泥捏成多个圆饼，大小方便食用即可。依次蘸取面粉、蛋液、面包糠，裹匀面衣。将裹好面衣的圆饼下锅炸，注意油温控制在 170 摄氏度，炸至表面酥脆即可出锅。制作卷心菜沙拉，将所需用料放入碗中，拌匀即可。装盘，在可乐饼旁边放上卷心菜沙拉、可乐饼蘸汁和黄芥末酱。

Beef and potato croquettes

[Makes 20]

4 potatoes (500g)
1 onion (200g)
200g beef trimmings
salt and pepper
--- to taste
30g butter

[coating]
flour, beaten eggs, breadcrumbs

vegetable oil
for deep-flying

[coleslaw]
200g shredded cabbage
1 tbsp lemon juice
2 tbsp olive oil
salt and pepper

[sauce]
3 tbsp *tonkatsu* sauce
2 tbsp tomato ketchup

mustard

1. Peel and cut each potato into 4-6 pieces. Soak them in water and drain. Line a microwave-resistant bowl with some paper towel and add the potatoes. Cover loosely with plastic wrap and microwave for 5-6 minutes until tender.
2. Remove the plastic wrap and paper towel from the bowl. Mash the potatoes while they are hot.
   * If you want to boil in a pot instead of using a microwave, see page 107.
3. Cut the onion into 1cm square pieces. Cut the beef into 2cm square pieces. Heat the butter in the frying pan and stir-fry the beef. Add the onions and continue to stir-fry. Season with 1/2 tsp salt and a little pepper.
4. Add the beef mixture and juices to the mashed potatoes. Season to taste with salt and pepper and mix well.
5. Shape into balls. Coat with flour, beaten egg, and then breadcrumbs. Deep-fry at 170 °C until crispy. Mix the ingredients of the coleslaw, and put onto a plate along with the croquettes. Serve with sauce and mustard.

炸猪排
"Tonkatsu" (pork cutlets)

炸猪排是一道日式家常菜肴，只要将猪肉裹上面衣下锅油炸即可。每次做这道料理的时候，除了当天吃的那份，我都会多做一份放进冰箱冷冻保存。这样一来，如果亲朋好友临时来访，我就可以拿准备好的炸猪排来招待他们。炸猪排可是我家冰箱里的常备菜品。

另外，建议大家尝试制作一些大小不一的炸猪排，大块的、小块的都准备一些。如此一来，您可以根据当天的心情和菜品种类选用适当尺寸的炸猪排，非常方便。单品食用炸猪排的时候，大块的猪排会让人觉得非常过瘾。而炸猪排盖饭我则经常搭配一些能够一口吃掉的小块炸猪排，这样就能多吃一点酥脆多汁的面衣啦。

"Tonkatsu" is deep-fried breaded pork. When I cook "tonkatsu," I always make much more than my family can eat on that day and store the rest in the freezer for unexpected visits from friends.
If you prepare various sizes of "tonkatsu" from bite-size to large cutlets, they can be used for a variety of recipes according to your needs. To make "tonkatsu", I use a whole cutlet and place it boldly on a plate. But when I make "katsu-don", I sometimes use bite-sized "tonkatsu", because the extra breading absorbs the soup and makes it tastier.

炸猪排

[用料 4人份]

猪上脑肉 1千克
或厚度2厘米的肉块 4块
盐、胡椒 各少许

[面衣]
面粉、蛋液、面包糠 各适量

卷心菜 1/2个
煎炸油 适量
炸猪排酱汁 适量
黄芥末酱 适量
柠檬 适量

1. 卷心菜切细丝,放入凉水浸泡稍许,使其口感更为爽口。沥干水分后,放进塑料保鲜袋,放入冰箱冷藏,食用之前取出即可。
2. 猪肉切块,每块厚度约2厘米。为了防止油炸后猪肉回缩,建议在瘦肉和肥肉之间的肉筋上划几刀。撒上盐和胡椒。
3. 猪肉表面依次裹匀面粉、蛋液、面包糠。
4. 炸制猪排,注意油温控制在170摄氏度。猪排熟透且表皮炸至焦黄色后,捞出,沥干油。
5. 将猪排切成一口大小,装盘,旁边适量摆放卷心菜丝、炸猪排酱汁和黄芥末酱。也可搭配柠檬和盐食用。

"Tonkatsu" (pork cutlets)

[Serves 4]

1kg or 4 slices (2cm-thick) pork shoulder loin
salt and pepper

[coating]
flour, beaten eggs, breadcrumbs

1/2 head cabbage
oil for deep-frying
tonkatsu sauce
Japanese mustard
lemon

1. Shred the cabbage and soak in ice water to make it crispy. Drain, put it in a plastic bag, and chill in the refrigerator until serving.
2. Slice the pork into about 2cm-thick slices. Slash the sinew running between the fat and lean tissue with a knife in order to prevent it from shrinking when deep-fried. Season with salt and pepper.
3. Dust the pork slices with flour, dip them in beaten eggs, and coat them with breadcrumbs.
4. Heat the oil to 170°C. Deep-fry the pork until golden brown on the outside and cooked through. Drain off excess oil.
5. Cut the fried pork into bite-sized pieces. Serve with the cabbage, your favorite sauce, and Japanese mustard. It's also tasty with just lemon and salt.

味噌青花鱼

Mackerel simmered in miso

　　典型的日式家庭菜肴。本食谱中使用的味噌口味清淡,除了老搭档米饭之外,搭配面包也别有一番风味。我想喝葡萄酒的时候,就会做一份味噌青花鱼,再搭配几片蒜香面包,妙哉妙哉!青花鱼建议切小块,这样吃到最后也不会腻。

This is one of the most popular and basic dishes in Japanese home cooking. However, my version is cooked with light miso sauce, which goes well with bread as well as with white steamed rice. When I feel like drinking wine, I lay slices of garlic bread along with this on my plate.
I like to cut the mackerel into small pieces rather than large ones. That way, I can savor each bite.

味噌青花鱼

[用料4人份]

青花鱼1条（净重400克）

生姜1块

酒1/2杯（100毫升）

水1/2杯（100毫升）

砂糖3大匙

酱油1大匙

甜料酒4大匙

味噌5~6大匙

1. 三片分割①青花鱼。切除鱼头（a），用刀尖划破鱼肚，掏出内脏。仔细清洗鱼肉后，用厨房纸巾擦干水分。水平持刀，从鱼尾侧入刀，沿着鱼脊骨横向划动刀具，片下鱼肉（b），再将鱼翻过来，按相同步骤片下另一面的鱼肉。剔除鱼脊骨（c）。稍倾斜菜刀，将上、下两块鱼肉各切成3等份。

   *三片分割中的"三片"指的是上、下2块鱼肉和中间1条鱼脊骨（d）。

2. 生姜去皮，切薄片。

3. 锅中倒入酒和水，开火，酒精挥发后加入砂糖、酱油、甜料酒、味噌，混合均匀。煮沸后，并排放入青花鱼块。放入姜片，汤汁再次煮开后盖上落盖，转小火，煮10~15分钟。

4. 煮到汤汁收少且变得黏稠时即可关火。鱼肉盛盘后，将所有汤汁浇在鱼肉上。

①三片分割：指将鱼身片成上、中、下三部分的日料常见切鱼技法。主要适用于青花鱼、沙丁鱼、竹荚鱼等纺锤形鱼类。

(a)

(b)

(c)

Mackerel simmered in miso

[Serves 4]

1 mackerel
(400g in fillets)
1 knob ginger
1/2 cup(100ml) sake
1/2 cup(100ml) water
3 tbsp sugar
1 tbsp soy sauce
4 tbsp mirin
5-6 tbsp miso

1. First, fillet the mackerel into three pieces: Cut off the head(a). Slice the stomach open with the tip of the knife and remove the innards. Wash thoroughly and wipe dry with a paper towel. Then, keeping the knife flat, cut horizontally from the tail along the side of the spine to loosen the top fillet(b). Turn it over and repeat on the other side. Shave off the bones around the stomach(c). Cut each fillet into three equal pieces in *sogi-giri*(diagonal-cut).
   * "Fillet a fish into three pieces" means two pieces of flesh and one of bone.(d)
2. Peel the ginger and thinly slice it.
3. Put the sake and water in a pan and turn on the heat. When the alcohol has boiled off, add sugar, soy sauce, mirin and miso into the pan and mix. When it comes to a boil, place the mackerel in the pan. Add the ginger. When the sauce comes to a boil again, put a drop-lid on and simmer for 10-15 minutes over low heat.
4. When the sauce is reduced and becomes syrupy, it is done. Arrange on a plate and pour the sauce on top with ginger slices.

(d)

香煮银鳕鱼

Aromatic stewed sablefish

大家可能觉得烹饪日式煮鱼有点难，但这道鱼料理非常简单，建议大家一定要尝试一下。只要掌握了调味汁的搭配比例，就可以用同样的方法烹饪其他品种的鱼。在苦菊和裙带菜上浇一些热乎的汤汁，您会收获一种意想不到的美味。

People often seem to think that cooking *nizakana* (simmering fish with broth) is a bit difficult compared to other Japanese dishes. But this is a very easy recipe. So, I'd like you to try it. Once you learn the balance of the ingredients for the broth, you can cook it with other fish as well.
The hot broth poured over raw *shungiku* and wakame turns this into an unexpected delicacy.

香煮银鳕鱼

[用料4人份]

银鳕鱼 4～5块（450～500克）

[调味汁]

酱油 4大匙

砂糖 2大匙

酒 4大匙

甜料酒 4大匙

味噌 1小匙

豆瓣酱 1～2小匙

蒜末 1大匙

生姜末 1大匙

葱末 1/2根葱的量

苦菊 适量

泡发好的裙带菜 适量

1. 摘下苦菊的叶子，将苦菊叶放入冰水稍微浸泡一会儿，使其口感更为爽脆。捞出，沥干水分备用。
2. 将泡发好的裙带菜切至适当大小，方便食用即可。把裙带菜和苦菊一起装盘，放入冰箱冷藏。
3. 擦干银鳕鱼的水分，每块都对半切开。
4. 锅中倒入调味汁，煮沸后放入银鳕鱼，注意鱼片之间不要重叠。
5. 再次沸腾后，放入蒜末、生姜末、葱末，盖上落盖，转小火，煮炖约10分钟后关火。
6. 装盘，将银鳕鱼放在步骤2的盘中，淋上煮鱼的汤汁。

Aromatic stewed sablefish

[Serves 4]

4-5 fillets sablefish
 (450～500g)

[seasonings]
4 tbsp soy sauce
2 tbsp sugar
4 tbsp sake
4 tbsp mirin
1 tsp miso
1-2 tsp *To-Ban-Jan*

1 tbsp minced garlic
1 tbsp minced ginger
1/2 Japanese leek, chopped
shungiku
 (chrysanthemum greens),
wakame seaweed
 --- to serve

1. Separate the leaves from the stems of the *shungiku*, and crisp the leaves by soaking in ice water, and drain well.
2. Cut the wakame into bite-sized pieces. Mix the *shungiku* and wakame, and put it on a serving plate. Cool it in a refrigerator
3. Pat dry the sablefish. Cut each fillet in half.
4. Combine the seasonings in a pan, and bring it to a boil. Add the fish, making sure you don't put the pieces on top of each other.
5. When it comes to a boil again, add the garlic, ginger, and leek. Put a drop-lid on and simmer for about 10 minutes over low heat.
6. Place the sablefish on top of the *shungiku* and wakame, and pour the piping hot sauce on top.

鲑鱼鲜虾丸

Salmon and shrimp *tsukune* meatballs

　　每次去国外,我都会惊讶于鲑鱼、鲜虾爱好者的庞大数量。这是我之前在国外的时候想出来的一个食谱。也可以用成品来制作烤串或汉堡,怎么吃都很美味。

Every time I go abroad, I am always amazed at how much people love salmon and shrimp. This is a recipe I came up with when I was abroad. They are also tasty in skewers like yakitori or as patties for hamburgers.

鲑鱼鲜虾丸

[用料 约10个]

生鲑鱼 3块（300克）
鲜虾仁 10只（200克）
洋葱 1/2个（100克）
酒 1大匙
盐、胡椒 各少许
色拉油 少许
酸橘（一切为二）适量
山椒粉 适量

[柠檬酱油]
甜料酒 1/2杯（100毫升）
酱油 1/2杯（100毫升）
柠檬汁 4大匙
昆布（长约5厘米）1片

[甜辣酱汁]
酱油 1/4杯（50毫升）
甜料酒 1/4杯（50毫升）
砂糖 2大匙

1. 鲑鱼去皮去骨，切块后用刀拍打。
2. 虾仁洗净，去虾线。切块后用刀拍打。尾部拍成泥，躯干部分不用拍成泥，以丰富口感。
3. 洋葱切小丁，边长7~8毫米。
4. 将拍打完的鲑鱼和虾仁放入碗中，加酒、盐、胡椒，混合均匀。加洋葱，再次拌匀。
5. 用步骤4的材料捏制丸子，每个丸子直径约5厘米。平底锅中倒入适量色拉油，油热后将丸子煎熟。
6. 制作柠檬酱油。用清水洗净昆布，擦干水分。锅中倒入甜料酒，开火，煮开后转小火继续煮制2~3分钟，关火，倒入碗中，再向碗中加入酱油、柠檬汁、昆布，放进冰箱冷藏。
7. 制作甜辣酱汁。把所需调味料倒入小锅中，开火，煮开后转小火继续煮约5分钟，至酱汁稍微变浓稠即可。
8. 将丸子盛入盘中，旁边放适量酸橘、山椒粉、柠檬酱油、甜辣酱汁。

Salmon and shrimp *tsukune* meatballs

[Makes 10]

3 salmon fillets (300g)
10 shelled shrimp (200g)
1/2 onion (100g)
1 tbsp sake
salt and pepper
vegetable oil
sudachi (halved)
--- to serve
sansho powder

[lemon *ponzu* sauce]
1/2 cup (100ml) mirin
1/2 cup (100ml) soy sauce
4 tbsp lemon juice
5cm-square-piece kombu kelp

[salty-sweet sauce]
1/4 cup (50ml) soy sauce
1/4 cup (50ml) mirin
2 tbsp suger

1. Remove the skin and bones from the salmon. Chop it coarsely first and then mince it.
2. Wash the shrimp and devein them. Chop it coarsely first and then mince it. Mince the tail completely and mince the body rather coarsely so that the texture remains.
3. Cut the onion into 7-8mm squares.
4. Put the salmon and shrimp in a bowl, add the sake, salt, and pepper, and mix. Add the onion and mix.
5. Shape the mixture into 5cm rounds. Pan-fry them until they are cooked through.
6. Make lemon *ponzu* sauce: Rinse the kombu lightly with water and wipe it dry. Put the mirin in a pan and bring it to a boil. Turn down the heat and simmer for a couple of minutes. Transfer the mirin to a bowl and add the soy sauce, lemon juice, and kombu. Keep in the refrigerator.
7. Make the salty-sweet sauce: Combine the ingredients in a small pan, and heat. When it comes to a boil, turn down the heat to low heat and let it simmer for about 5 minutes until it thickens a little.
8. Serve the *tsukune* with sudachi, *sansho* powder, lemon *ponzu* sauce and salty-sweet sauce.

日式腌渍鲑鱼

Deep-fried salmon marinated in *nanbanzu*

　这道料理是将油炸过的食材和未炸过的红辣椒等食材一起泡入混合醋汁中腌制而成的食物。我想要多吃一点蔬菜，所以这道食谱里也加了很多蔬菜。

　这道菜品适用于一年四季，但不同季节我也会相应添加不同的果汁来调味。例如，冬天用柚子，夏天用酸橘或柠檬。加入一些香气四溢的时令柑橘类水果会使菜肴更加美味，建议大家也尝试一下。

Nanbanzuke is a dish in which ingredients are deep-fried and then marinated in vinegared sauce with red peppers. Because I want to eat lots of vegetables, I use plenty of them in this recipe.
I cook this dish throughout the year, and use seasonal citrus fruits, such as yuzu in winter and sudachi or lemons in summer, to enjoy their fragrance.

日式腌渍鲑鱼

[用料 1 人份]

生鲑鱼片 4 块（400 克）
胡萝卜 1 小根（100 克）
洋葱 1/2 个（100 克）
芹菜 1/2 根（净重 80 克）
生姜 1 小块
红辣椒圈 2 根辣椒的量
柚子 适量
面粉 3 大匙
盐、胡椒 各少许
煎炸油 适量

[混合醋汁]

日式高汤 1 杯（200 毫升）
淡口酱油 3 大匙
醋 3/4 杯（150 毫升）
砂糖 4 大匙
盐 少许

1. 胡萝卜切丝，长 4～5 厘米。
2. 洋葱切薄片。
3. 芹菜去筋，切丝，长 4～5 厘米。
4. 生姜切丝。
5. 制作混合醋汁。将日式高汤、淡口酱油、醋、砂糖、盐倒入碗中，搅拌均匀。

*可根据个人喜好，添加 1 大匙柚子、柠檬、酸橘等柑橘类水果的果汁。

6. 将鲑鱼切至适当大小，方便食用即可。撒盐和胡椒，裹满面粉，下锅油炸，注意油温控制在 170～180 摄氏度。趁热将鲑鱼放入混合醋汁中浸泡，加入切好的胡萝卜、洋葱、芹菜、生姜和红辣椒圈后，在最上面放几块柚子皮提味。用保鲜膜包住碗口，放入冰箱冷藏入味。

Deep-fried salmon marinated in *nanbanzu*

[Serves 4]

---

4 salmon fillets (400g)
1 small carrot (100g)
1/2 onion (100g)
1/2 stalk celery (80g)
1 small knob ginger
2 red chili peppers, chopped
yuzu
3 tbsp flour
salt and pepper
vegetable oil
  for deep-frying

[*nanbanzu* sauce]
1 cup (200ml) dashi
3 tbsp light soy sauce
3/4 cup (150ml) vinegar
4 tbsp sugar
salt

1. Cut the carrot into 4-5cm thin strips.
2. Slice the onion.
3. String the celery and cut it into 4-5cm thin strips.
4. Cut the ginger into thin strips.
5. Combine the dashi, light soy sauce, vinegar, sugar, and salt in a bowl, and mix them to make the *nanbanzu* sauce.
   * You can add 1 tbsp juice of citrus such as yuzu, lemon, or sudachi if preferred.
6. Cut the salmon into bite-sized pieces. Sprinkle salt and pepper over the salmon and cover the salmon completely with flour. Deep-fry the salmon in oil at 170-180°C. Marinate the salmon in *nanbanzu* while they are hot. Add the carrot, onion, celery, ginger, red chili pepper and top it with yuzu zest for fragrance. Cover with plastic wrap and put in the refrigerator. Let it stand for a while to marinate.

其实对于不太大的虾来说，与其直接一只一只下锅油炸，不如将几只团在一起做成虾排。这样炸制而成的虾排，在口感、外观、美味程度、嚼劲上都会有新的突破。注意虾不要炸过头。我们家聚会的时候经常做这道料理，刚出锅热乎乎的时候非常好吃。

Small shrimp better satisfy your hunger and have a more appealing look, taste and texture when they are deep-fried together in a cutlet than fried separately. Be careful not to overcook the shrimp. I always serve this dish sizzing hot at my home parties.

炸虾排

[用料8个]

虾 24只
面粉 6大匙
蛋液 1个鸡蛋的量
水 1大匙
面包糠 适量

[塔塔酱[1]]
切碎的酸黄瓜 3大匙
切碎的煮鸡蛋 2个鸡蛋的量
洋葱碎 1/4个洋葱的量（3大匙）
蛋黄酱 1杯（200毫升）
牛奶 1～2大匙
黄芥末 少许

盐、胡椒 各适量
煎炸油 适量
柠檬 适量

1. 调制塔塔酱。碗中依次放入蛋黄酱、牛奶、黄芥末，搅拌均匀。加切碎的酸黄瓜、煮鸡蛋和洋葱碎，拌匀后，加适量盐和胡椒调味。
2. 碗中放入面粉、打匀的蛋液、水，搅拌均匀至糊状。
3. 生虾剥壳、去尾，用牙签挑去虾线。将处理好的虾切成上下两半。将每六块虾肉团成一个饼状虾排，加适量盐和胡椒调味。
4. 用平铲等工具铲起步骤3制作的虾排，在其表面涂抹步骤2调制的面糊，裹适量面包糠。
5. 下油锅将虾排彻底炸透，注意油温控制在180～190摄氏度。
6. 装盘，搭配塔塔酱、其他喜欢的酱汁、柠檬食用。

---

[1] 塔塔酱　英文写作"Tartar sauce"，小称鞑靼沙司。

Ebikatsu (shrimp cutlets)

[Makes 8]

24 shrimp
6 tbsp flour
1 egg, beaten
1 tbsp water
breadcrumbs

[tartar sauce]
3 tbsp pickled cucumber, finely chopped
2 boiled eggs, finely chopped
1/4 finely chopped onion (3 tbsp)
1 cup (200ml) mayonnaise
1-2 tbsp milk
Japanese mustard

salt and pepper
vegetable oil for deep-frying
lemon

1. Make the tartar sauce: Put the mayonnaise in a bowl, and dilute it with the milk. Add a little Japanese mustard, and mix. Mix in the pickled cucumber, egg, and onion. Season to taste with salt and pepper.
2. Mix the flour, beaten egg, and 1 tbsp water until the mixture becomes paste-like.
3. Remove the shells and tails from the shrimp. Devein them with a toothpick. Cut the shrimp in half. Use 6 pieces per portion to make round, tight, flat patties, and season with salt and pepper.
4. Scoop the patties using a turner and spread the paste made in step 2 over the surface. Then coat the patties with breadcrumbs.
5. Deep-fry the patties in oil at 180-190°C until they are cooked through.
6. Serve with your favorite sauce, tartar sauce, and lemon.

樱花 | Sakura

樱 花

春天的樱花,是我最为喜爱的花。

日本国内自不必说,现如今樱花在世界范围也备受喜爱。

每当天气一天一天地变暖,春天一天一天地走近的时候,

我都翘首以盼,期待着樱花的盛开。

樱花开花不易,但花谢只需十几天,

干净利落又短暂脆弱。

不知为何,樱花有一种让人平静下来的力量。

每年樱花季,我都会在玄关和房间里放上几枝樱花,

也会轻摘一朵樱花,放进一杯酒、一盏茶中。花瓣漂浮在水面上,赏心悦目。

有时,我也会折一小枝樱花用作筷架。

另外,我也收集了很多樱花主题的小物件,比如器具和布艺品。

春天是我一年四季中最喜欢的季节。

每年这个季节,市面上都能买到各种美味的鱼和新鲜的蔬菜,

好闻的叶菜更能激起我下厨的欲望。

生在日本,我很高兴。

在这食材丰饶、温暖优美的季节里,定要不负时光,不负每一天。

Sakura

Spring in Japan reminds me of sakura (cherry blossoms),
my favorite flower.
Sakura are now widely loved by many people in the world,
and not only by Japanese people.
When it gets warmer and spring feels like it's just around the corner,
I am anxious for the news that cherry-blossoms are in bloom.
Once cherry blossoms have fully opened,
the petals soon fall to the ground.
They only last for about ten days after the buds have bloomed.
They are gracious and, though very fragile,
the blossoms somehow soothe people's minds.
During cherry-blossom season,
I put some cherry blossoms in the hallway or
in the rooms of my home for decoration, or pick some petals
and set them nicely afloat on a cup of Japanese sake or green tea.
Sometimes, I pick a tiny twig off a sakura tree
and place it on the table as a chopstick rest.
I now have many kinds of tableware or tablecloths with
cherry-blossom motifs on them.
Spring is my favorite season of the year.
Many delicious fish and vegetables become available in the market.
When I see sweet-smelling herbs or fresh leafy vegetables,
it arouses in me a feeling of enthusiasm for cooking.
I am happy to be born in Japan, and I am enjoying living here.
I like to cherish every moment of this rich
and generous season of the year.

本书中使用的主要调味品
Basic ingredients used in this book

酱油　Soy sauce

极为常见的调味品,以黄豆为原料。带有咸味,滋味独特,是日式菜肴中不可或缺的调味品。酱油味道鲜美,香味浓郁,用途广泛,既可以直接作为食材蘸料使用,也可以用来烹饪炖菜、调制酱汁等。酱油种类丰富,可粗略分为浓口和淡口两种,平时提到的酱油一般是指浓口酱油。相较于淡口酱油,浓口酱油颜色较深但含盐量较低。

The soy-based sauce essential for Japanese recipes contains salt and has a uniquely savory taste. It is widely used as dipping sauce, for seasoning simmered food, and for adding flavor to Japanese-style sauces. There are various kinds of soy sauce, with regular soy sauce (*koikuchi*) offering darker color and less salt compared with light soy sauce (*usukuchi*).

淡口酱油　Light soy sauce

色泽和香味较淡、含盐量较高的酱油。能够尽可能保持食材本身的味道和颜色,是日本关西地区主要使用的酱油。常用于煮制浅色蔬菜和白肉鱼,不希望汤汁颜色过深的时候也可使用。所有酱油中含盐量最高的便是淡口酱油。

This soy sauce, which is favored in the Kansai area, has moderate color and flavor but is saltier than regular soy sauce. Light soy sauce is in fact the saltiest of all soy sauce varieties. Harmonizing with the natural flavor and color of ingredients, it is used for simmering pale-colored vegetables or white fish, and for seasoning clear soups.

谷物醋　*Kokumotsu-su* (Rice and grain vinegar)

日本家庭中最常见的醋。以大米、小麦、酒糟、玉米等为原料制成。特点是酸味清淡,不仅可用于日式菜肴,也广泛应用于西餐和中餐里。酸度比葡萄酒醋低了近1.2%。

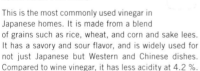

This is the most commonly used vinegar in Japanese homes. It is made from a blend of grains such as rice, wheat, and corn and sake lees. It has a savory and sour flavor, and is widely used for not just Japanese but Western and Chinese dishes. Compared to wine vinegar, it has less acidity at 4.2 %.

料理酒　*Ryori-shu* (Cooking sake)

烹饪用酒。与用来喝的日本酒不同,料理酒中含有酸味且纯度不高,但可以去除鱼、肉类的腥膻味,增加菜肴的香气。有些料理酒中添加了盐、甜味剂等。亦可使用普通的日本清酒代替料理酒。

This is sake used for cooking. Unlike the sake for drinking, the cooking sake is a little sour and has a bit of an undesirable taste but it helps to remove the unpleasant smell of fish and meat, and gives deeper flavor to dishes. Some products have added salt or sugar etc. You can substitute it with ordinary sake for drinking.

甜料酒　Mirin

以蒸熟的糯米、米曲和烧酒为原料酿制而成的甜酒。主要作用是为料理增加甜味和光泽度。与砂糖不同,甜料酒的甜味更加温和,也更为鲜美。另外,由于含有酒精成分,所以甜料酒既能够有效地增加食材韧性、防止食材煮烂,又能够去除鱼、肉类的腥膻味。

Sweet alcoholic liquid produced from distilled spirits, steamed glutinous rice, and malted rice. Mirin adds sweetness and luster to ingredients. Compared with sugar, mirin offers a milder sweetness and flavor. The alcohol contained in the mirin protects simmered ingredients from crumbling and also removes the smell of meat and fish.

味噌 Miso

以黄豆为主要原料，加入米、麦发酵而成，是日本自古以来一直食用的传统调味品。不同地区的味噌在色、香、味上各有不同，据说日本境内有上百种味噌。本书中若无特殊说明，指的都是偏黄色的浅色米味噌。

Traditional Japanese seasoning produced from fermented soybeans, rice, and wheat. There are several hundred regional variations in taste, color, and flavor of miso paste. In this book, I use rice miso (komemiso), which has a yellowish color, except in cases where I suggest using other specific miso pastes.

上白糖 Johakuto (Refined white sugar)

日本家庭中最常用的砂糖。本书食谱中提到的砂糖基本上指的都是上白糖。它是日本特有的砂糖品种，与细砂糖相比结品颗粒细小，性状湿润。其特点是甜味清淡爽口，易溶于水。

This is the sugar most commonly used in Japanese homes. If the recipe says "sugar", it usually implies this sugar. It is unique to Japan and has finer crystals than granulated sugar and contains some moisture. It has no strong flavor and dissolves easily.

鸡精 Granulated Chinese chicken soup stock

以鸡汤提取物为底料，添加蔬菜提取物调味。主要应用于中餐的速食调料。可在煲汤、炒菜时适量使用。

Ready-made seasoning produced from the essence of chicken soup stock and flavored with vegetable essence. It is mainly used for Chinese recipes such as soups and sautéed dishes.

炸猪排酱汁 Tonkatsu sauce

炸猪排专用酱汁，模仿英国伍斯特沙司制成。但与伍斯特沙司相比，炸猪排酱汁甜味更突出，酱汁更浓稠。

This Japanese sauce is mainly used for deep-fried pork (*tonkatsu*) and it imitates England's famous Worcester sauce. Although it is a kind of Worcester sauce, *ton katsu* sauce is sweeter and thicker than other Japanese-style Worcester sauces.

芝麻油 Sesame oil

芝麻经焙炒后压榨而成，色如琥珀，香味独特。

Made from compressed sesame seeds. Regular sesame oil is made from toasted sesame seeds and is brown in color with a strong flavor.

豆瓣酱 To-Ban-Jan

中式调味料，常用于川菜。含有辣椒，味道鲜美，可给菜肴增酸添香、增添辣味，炒菜、炖菜、勾芡时均可使用。

Chinese chili paste mainly used for Sichuan-style recipes. *To-Ban-Jan* is used to add spiciness and a savory taste to sautéed dishes, simmered dishes, and thick liquid starch sauces.

片栗粉（土豆淀粉）
Potato starch (*Katakuri ko*)

日文中"片栗"意为"猪芽花"，片栗粉原指由猪芽花根茎制成的淀粉，也因此而得名。但现在市面上常见的其实是土豆淀粉。常用于制作油炸食物的面衣以及用来勾芡。用片栗粉勾芡时，为防止结块，需提前用冷水调匀后再加入菜肴，使汤汁变得浓稠。

Traditional Japanese *katakuri ko* used to be made from the root of *katakuri*, and was named after the plant. Today, starch made from potato is available as a substitute for *katakuri ko*. Potato starch is used for deep-fried batter and as a thickener of soups and sauces. When using it as a thickener, dissolve it in water before use so that it does not become lumpy.

2

VEGETABLES

蔬菜料理

胡萝卜金枪鱼沙拉

Carrot and tuna salad

  我第一次做这道菜,是因为当时家里有好多胡萝卜,想用手里现有的食材尝试开发新菜品。一晃三十多年过去了,这道沙拉依然很受欢迎。

  本料理中使用的胡萝卜既不是生的也不是水煮的,而是用微波炉加热过的。另外,洋葱末和蒜末是这道菜的点睛之笔。

When I had a lot of carrots left unused I came up with this salad recipe, using other ingredients that I had in my kitchen. I first introduced this recipe more than 30 years ago, but it is still quite popular.
The carrots are not boiled, nor eaten raw. They are microwaved. The key to this recipe is to finely chop the onion and garlic.

胡萝卜金枪鱼沙拉

[用料4人份]

胡萝卜 250克

洋葱末 2大匙

蒜末 1小匙

色拉油或橄榄油 1大匙

金枪鱼罐头 1/2小罐（30克）

[调味汁]

白葡萄酒醋 1大匙

黄芥末粒 1大匙

柠檬汁 1大匙

盐、胡椒 各适量

1. 胡萝卜切丝，每根长5～6厘米。将胡萝卜丝放入耐热容器，加洋葱末、蒜末、色拉油拌匀。
2. 用保鲜膜轻轻地包住容器口，用微波炉（600瓦）加热1分10秒～1分20秒。
3. 取出容器，取下保鲜膜。轻轻搅拌，依次放入沥干油分的金枪鱼肉和调味汁，拌匀。
4. 放入冰箱冷藏入味。

Carrot and tuna salad

[Serves 4]

250g carrot
2 tbsp onion, finely chopped
1 tsp garlic, finely chopped
1 tbsp vegetable oil
  or olive oil
1/2 small can of tuna(30g)

[dressing]
1 tbsp white wine vinegar
1 tbsp grain mustard
1 tbsp lemon juice
salt and pepper --- to taste

1. Cut the carrot into 5-6cm thin strips. Put them in a microwave-resistant bowl and mix in the onion, garlic and oil.
2. Cover it loosely with plastic wrap and microwave at 600W for 1 minute and 10 to 20 seconds.
3. Remove the bowl from the microwave and remove the wrap. Mix lightly and add the drained tuna and the dressing ingredients. Mix together.
4. Chill in the refrigerator to allow the dressing to soak into the vegetables.

土豆沙拉

Potato salad

有各种各样的土豆沙拉食谱,但我觉得这个方子是最简单也最常见的家的味道。本食谱用料简单,入手容易,对制作时间和制作地点都没有限制。这道菜美味的秘诀是将黄瓜稍微切厚一点以丰富口感,洋葱泡水时间不要过久,以及加蛋黄酱搅拌的时候动作要轻。

There are many variations of potato salad recipes, but this one is the simplest, most basic taste enjoyed in home cooking across Japan. The ingredients are easily available in any country, so you can make it any time, anywhere. The tip to make it delicious is to slice the cucumbers slightly thick for crispy texture, not to soak the sliced onions in the water for too long, and to fold in the mayonnaise gently.

土豆沙拉

[用料4人份]

土豆4~5个（450克）
黄瓜1根
洋葱（小）1/2个
火腿2片
蛋黄酱5~6大匙
盐、胡椒 各适量

1. 土豆去皮，切成4~6等份，泡入清水中以去除部分淀粉，捞出，沥干水分。在耐热容器内铺上厨房纸巾，放入土豆块。用保鲜膜包住容器口，注意不要包得太紧，用微波炉（600瓦）加热5~6分钟至土豆变软。

2. 取下保鲜膜，取出厨房纸巾，趁热捣碎土豆，制成土豆泥，自然放凉。

   *如需用锅具煮制土豆，请按以下步骤操作：仔细洗净土豆，将未削皮的整个土豆放入锅中，向锅中倒入清水，水量刚好没过土豆即可。盖上锅盖，开火加热。水开后转小火继续煮20~25分钟至土豆变软，捞出土豆，趁热剥去土豆皮。将土豆放入碗中，捣碎。

3. 黄瓜切片，放盐后静置5分钟。黄瓜变软后挤干多余水分。

4. 洋葱切片，泡水5分钟后挤去多余水分。

5. 火腿切条。先将火腿片对半切开，再将其切成1厘米宽的小条。

6. 向土豆泥中放入黄瓜、洋葱、火腿，搅拌均匀。加蛋黄酱，再次拌匀后，用盐和胡椒调味。

Potato salad

[Serves 4]

4-5 potatoes(450g)
1 cucumber
1/2 small onion
2 ham slices
5-6 tbsp mayonnaise
salt and pepper

1. Peel and cut each potato into 4-6 pieces. Soak them in water and drain. Line a microwave-resistant bowl with some paper towel and add the potatoes. Cover loosely with plastic wrap and microwave for 5-6 minutes until tender.
2. Remove the plastic wrap and paper towel from the bowl. Mash the potatoes while they are hot and let it cool.
   * To boil in a pot, follow these instructions: Thoroughly wash unpeeled whole potatoes, put them in a pot, cover with water, put on a lid and turn on the heat. When it comes to a boil, turn the heat down to low, and cook for 20-25 minutes until tender. Peel the skin and put them in a bowl while they are still hot and mash them.
3. Thinly slice the cucumber. Toss with a little salt and let it stand for 5 minutes. When it starts to soften, squeeze the excessive moisture out.
4. Thinly slice the onion, soak in water for 5 minutes, and squeeze.
5. Cut the slices of ham in half and then cut into 1cm-wide strips.
6. Add the cucumber, onion and ham to the mashed potatoes, and mix together. Add the mayonnaise and mix again. Season with salt and pepper to taste.

果仁菠菜

Spinach with peanut sauce

在世界各地都可以买到菠菜和花生酱,我在出国的时候经常做这道菜。这个食谱很简单,但要做得好吃需要注意菠菜焯水的火候,确保挤干菠菜中的水分。

This dish is made with spinach and peanut butter, both of which are readily available in any country. Therefore, I often cook this when I am overseas. It is a very simple and easy recipe. Be sure not to overcook the spinach and to thoroughly squeeze the excess water out. This is necessary to make the dish excellent.

果仁菠菜

[用料 4人份]

菠菜 200克
花生酱（含糖型）3大匙
砂糖 1小匙
酱油 1小匙
甜料酒 2小匙
盐 适量
花生碎 适量

1. 碗中放入花生酱、砂糖、酱油、甜料酒，拌匀。
2. 将菠菜切段，每段长度约为5厘米。将菠菜茎叶分开。锅中加水烧开，加少许盐，先放茎，再放叶，将菠菜快速焯水。捞出菠菜放入冷水，用漏勺捞出，仔细挤干水分。
3. 将挤干水分的菠菜放入步骤1调制的酱汁中。尝下咸淡，如有必要可少量加盐。
4. 装盘，撒上一些花生碎。

Spinach with peanut sauce

[Serves 4]

200g spinach
3 tbsp peanut butter (sweetened type)
1 tsp sugar
1 tsp soy sauce
2 tsp mirin
salt
peanuts

1. Mix the peanut butter, sugar, soy sauce, and mirin in a bowl.
2. Cut the spinach into 5cm-long pieces and separate the leaves from the stems. Blanch the stems first in boiling water with a pinch of salt, and then add the leaves. Plunge in cold water and then drain well. Squeeze out the excess water.
3. Add the spinach to the peanut sauce and mix. Check the taste and add a little salt if necessary.
4. Serve the spinach onto a plate and sprinkle coarsely chopped peanuts on top.

焯拌番茄

Tomato *ohitashi*

一般来讲，提到焯拌菜大家的第一反应都是拌青菜。但我经常用焯拌菜的方法处理各种食材，也尝试过秋葵、菌菇类、山药泥等版本。焯拌菜的方法适用于制作各种蔬菜，大家可以使用自己国家的常见蔬菜随意尝试。

Ohitashi is often made with leafy vegetables, but I make it with all kinds of ingredients, such as okra, mushroom, grated yam, among others. This recipe works for any vegetable, so please feel free to be creative with whatever ingredient comes in handy in your country.

焯拌番茄

[用料 易于制作的分量]

中号番茄 2 袋（400 克）
日式高汤 1+1/2 杯（300 毫升）
甜料酒 2 大匙
淡口酱油 2 大匙

1. 将日式高汤、甜料酒、淡口酱油倒入碗中，混合均匀。
2. 番茄去蒂，底部划一道斜口。
3. 锅中加水烧开，放入番茄快速焯水，迅速捞出番茄后将其放入冷水中，捞出，沥干水分，去皮。
4. 将番茄放入步骤1的调味汁中浸泡，用保鲜膜包住碗口，放入冰箱冷藏入味。

Tomato *ohitashi*

[Ingredients]

2 packs middle-sized tomatoes(400g)
1+1/2cups(300ml) dashi
2 tbsp mirin
2 tbsp light soy sauce

1. Combine dashi, mirin and light soy sauce in a container.
2. Remove the stems of the tomatoes. Slice a shallow cut into the bottom of the tomatoes.
3. Place the tomatoes into boiling water. Put them into a bowl of cold water immediately, drain them, and peel off the skins.
4. Soak the tomatoes in the dashi mixture prepared in step 1. Cover the container with plastic wrap. Chill in the refrigerator to let the tomatoes absorb the flavors.

卷心菜沙拉

Coleslaw

我很喜欢以酸味为主、甜味为辅的卷心菜沙拉。蔬菜水分过多会影响沙拉的口感，仔细挤干卷心菜和胡萝卜中的水分后，再倒入调味汁为佳。

I like coleslaw that is sour and a bit sweet. It wouldn't taste good if it gets watery, so the key is to thoroughly squeeze the excess water from the shredded cabbage and carrot before mixing in the dressing.

卷 心 菜 沙 拉

[用料 易于制作的分量]

卷心菜4～5片（400克）
胡萝卜50克
盐1+1/2小匙

[调味汁]

蛋黄酱6大匙
醋3大匙
砂糖1小匙

盐、胡椒 各少许

1. 卷心菜切成片，每片边长约1厘米。放入碗中备用。
2. 胡萝卜去皮，切丁，每块边长约8毫米。放入碗中。加盐拌匀，放置10～15分钟。胡萝卜腌出水后，用棉布等工具仔细挤出多余水分。
3. 制作调味汁，将所需调料混合均匀即可。
4. 将卷心菜和胡萝卜放入碗中，倒入调味汁拌匀，加入盐和胡椒调味。

Coleslaw

[Ingredients]

4-5 leaves cabbage(400g)
50g carrot
1+1/2 tsp salt

[dressing]
6 tbsp mayonnaise
3 tbsp vinegar
1 tsp sugar

salt, pepper

1. Cut the cabbage into 1cm squares and put them into a bowl.
2. Peel the carrot, thinly slice it into 8mm squares and add them into the bowl. Mix in the salt, and leave for 10-15 minutes. When excess water comes out, squeeze it out thoroughly using a cheesecloth, etc.
3. Combine the ingredients for the dressing in a bowl.
4. Add the mixture of dressing to the bowl of cabbage and carrot, and mix well. Season with salt and pepper.

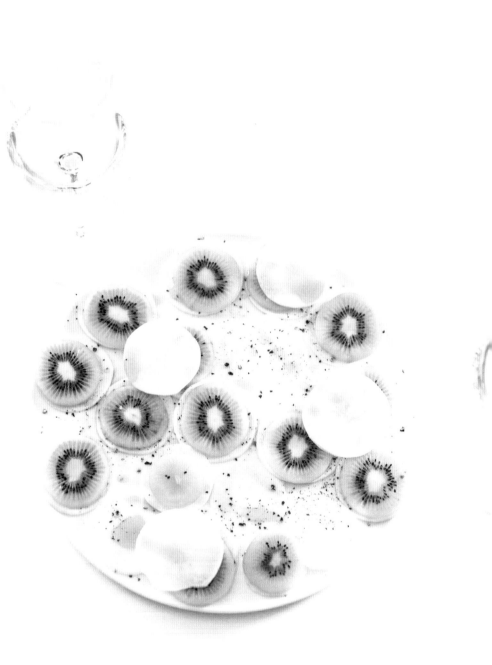

芜菁猕猴桃薄片沙拉

Turnip and kiwifruit carpaccio

这是一道典型的快手菜，5分钟便可以做好。提前切好食材并将其放入冰箱充分冷藏，之后只需装盘即可。掌握几道快手菜，家里来客的时候会很方便。当着朋友的面装盘的话，朋友也会觉得新奇有趣。

All you have to do is to cut the ingredients and chill them well in the fridge. All that is left is to arrange them on a plate. This simple dish is ready in just 5 minutes. Having several easy recipes like this one will come in handy when you have people over. I entertain my friends by arranging the food on the plate in front of them.

芜菁猕猴桃薄片沙拉

[用料4人份]

芜菁 2~3个
猕猴桃 1~2个
盐、胡椒 各少许
橄榄油 适量

1. 芜菁去茎去皮切圆片,切薄一点。
2. 猕猴桃去皮切圆片,切薄一点。
3. 先将芜菁放入碗中,撒盐和胡椒,再将猕猴桃置于其上。之后再将一些芜菁片点缀在猕猴桃片上面,不用盖满,随意堆叠即可。将整盘料理放入冰箱冷藏,食用时取出即可。
4. 淋一圈橄榄油后即可食用。

Turnip and kiwifruit carpaccio

[Serves 4]

2-3 turnips
1-2 kiwifruits
salt, pepper
olive oil

1. Cut the stems off the turnips, peel, and cut into thin round slices.
2. Peel the kiwifruits and cut them into thin round slices.
3. Place the turnip slices on a serving plate, sprinkle some salt and pepper. Place the kiwifruit slices on top of the layer of turnip. Add some turnip slices on top of the kiwifruit partially, and let it cool in the refrigerator until serving.
4. Sprinkle olive oil right before eating.

脆爽拍黄瓜

Easy pickled cucumber

简单美味的家常凉菜。即便桌上有很多珍馐美味，这道小菜也依然是我家人的最爱。这道菜刚做好的时候很好吃，放上两三天更加入味，也别有一番风味。建议大家每次可以多做一点，多吃几天。

This is a small side dish. But this dish is always the most popular among my family even when there are other delicacies on the table. This is good not just on the day it is made, but it gets tastier in the next few days as it absorbs more flavor. So make plenty at a time and enjoy !

脆爽拍黄瓜

[用料 4人份]

黄瓜 6 根
生姜 1 块
酱油 1/2 杯（100 毫升）
醋 1/2 杯（100 毫升）
砂糖 4 大匙
红辣椒圈 1～2 根辣椒的量

1. 将酱油、醋、砂糖倒入容器，混合均匀。
2. 把黄瓜两端切掉，用研磨棒或擀面杖等工具拍打黄瓜，至刚好产生裂缝的程度。每根黄瓜切成 4～6 等份。
3. 生姜切丝。
4. 把黄瓜放入塑料保鲜袋后，将步骤 1 的调料汁和红辣椒圈倒进袋中。放入冰箱冷藏至少 2～3 小时。
5. 食用之前放入姜丝提味。

Easy pickled cucumber

[Serves 4]

6 cucumbers
1 knob ginger
1/2 cup(100ml) soy sauce
1/2 cup(100ml) vinegar
4 tbsp sugar
1-2 red chili peppers, chopped

1. Combine the soy sauce, vinegar, and sugar.
2. Trim the ends of the cucumbers and strike them with a rolling pin to make cracks. Cut each cucumber into 4-6 pieces.
3. Cut the ginger into fine strips.
4. Put the cucumbers into a plastic bag and add the mixture prepared in step 1 and the red chili pepper. Let stand in the refrigerator for at least 2-3 hours.
5. Add the ginger strips right before serving to let the cucumbers absorb the flavor.

美味豆腐

"Gochiso-dofu" (decorated tofu)

我们可以将各种口感、味道不尽相同的美味食材切碎，置于这道美味豆腐上面。我一直都用冰箱里现有的食材制作配料。每次有客人临时来访，我都会用这道简单美味的菜品来招待他们。为了保证菜品色香味俱全，围住豆腐的厨房纸巾一定要比豆腐高2厘米，这样既可以吸收豆腐里多余的水分，又可以防止顶部配料溢出。另外，还要注意不要直接从豆腐正上方浇淋万能酱油，应该沿着豆腐边缘少量多次地进行操作。

This *gochiso-dofu* is served with all kinds of ingredients cut into small pieces that have different textures or tastes. Toppings on tofu are usually leftovers from my fridge. This is a very convenient and smart dish when we have unexpected guests at our home.

To keep the pleasant appearance of the toppings, it is important to pay attention to details such as wrapping the paper towel around the tofu to absorb excess water, making the edge of the towel extend 2cm above the tofu so the toppings don't fall off, and pouring the sauce little by little along the edge of the tofu.

美味豆腐

[用料 2~4人份]

绢豆腐 1块
火腿 2片
（可用烤牛肉、烤鸡肉、烤金枪鱼等食材代替）
生姜 20克
葱花 3大匙
青紫苏① 5片
花生、白芝麻 各适量

[万能酱油]
昆布（5厘米大小）1片
甜料酒 1/4杯（50毫升）
酱油 3/4杯（150毫升）

1. 制作万能酱油。轻轻地清洗昆布并用厨房纸巾擦干。将甜料酒倒入小锅中煮沸，再熬制1~2分钟后关火。趁热加入酱油和昆布。常温放置至少1小时后取出昆布。
2. 去除豆腐中的水分后将其置于容器中。用厨房纸巾包住豆腐侧面以吸收多余的水分，注意厨房纸巾应比豆腐高2厘米。放入冰箱，充分冷藏。
3. 准备配料。切好火腿、生姜，加入葱花、青紫苏、花生。
4. 将步骤3准备的配料置于冷藏后的豆腐上，撒上白芝麻，随后取走厨房纸巾。
5. 沿着豆腐边缘少量多次地浇淋万能酱油，注意不要直接从豆腐正上方浇，以防破坏料理装盘，影响美观。

---
① 青紫苏：日料中常见的香味蔬菜，香气清新，是开胃的食材。

"Gochiso-dofu" (decorated tofu)

[Serves 2-4]

1 pack soft tofu
2 slices ham
(Roast beef, roast chicken,
or canned tuna
can be substituted.)
20g ginger
3 tbsp chopped spring onion
5 shiso leaves (green perilla)
peanuts, white sesame seeds
--- to taste

[all-purpose soy sauce]
1 piece kombu kelp
(5cm square)
1/4 cup(50ml) mirin
3/4 cup(150ml) soy sauce

1. Make the all-purpose soy sauce: Rinse the kombu lightly, and wipe with a paper towel. Put the mirin in a small pan and bring to a boil. Boil it for 1-2 minutes, and turn off the heat. Add the soy sauce and kombu while still hot. Let it stand for more than 1 hour and then remove the kombu.
2. Drain the tofu and place on a serving plate. Wrap a folded paper towel around the sides of the tofu to absorb excess water. Make sure the edge of the paper towel extends 2cm above the tofu. Place in the refrigerator to chill the tofu thoroughly.
3. Chop up the toppings: ham, ginger, spring onion, shiso and peanuts.
4. Pile the toppings from step 3 on the chilled tofu. Sprinkle sesame seeds, and remove the paper towel.
5. Pour the all-purpose soy sauce little by little along the edge of the tofu so that it doesn't disturb the beautifully arranged toppings.

千层豆腐

Tofu lasagna

提到"千层",大家往往想到的都是意大利菜里的千层面。但经常出现在我家餐桌上的却是以豆腐或茄子为主料的千层豆腐。这道菜看上去很复杂,但是用生奶油替代白酱的话就会简单很多。一定要多放一些奶酪。

Lasagna is originally made of pasta. However, I use tofu or eggplant instead. You may think making lasagna is time-consuming, but with my recipe, you can easily cook it using heavy cream instead of white sauce. Please put plenty of cheese on top.

千层豆腐

[用料 4人份]

绢豆腐 2块
肉酱 2杯（400毫升）
(参照本书食谱制作或购买市售产品)
生奶油 1/4杯（50毫升）
奶酪（比萨用）150克
橄榄油 2大匙
盐、胡椒 各少许

1. 用厨房纸巾包住豆腐以吸收多余水分，放置约15分钟后，把每块豆腐切成5大片。
2. 向平底锅中倒入橄榄油，中火加热。将豆腐片放入锅中，撒盐和胡椒，将豆腐的两面煎至变色。
3. 烤箱预热至230摄氏度。
4. 在耐热容器底部薄薄地铺上一层肉酱，将豆腐片放在肉酱上，然后倒上一层生奶油。重复此步骤，最后在顶部撒上奶酪。
5. 放入烤箱，烤制15～20分钟。

肉 酱

[用料 4人份]

猪肉、牛肉混合肉馅 500克
培根 50克
洋葱 1个（200克）
胡萝卜（小）1/2根（50克）
芹菜 1/2根（50克）
蘑菇 1包（100克）
蒜末 1瓣蒜的量
橄榄油 2大匙
盐、胡椒 各少许
红葡萄酒 1/2杯（100毫升）
半冰沙司①（市售）1罐（290克）
番茄汁 1杯（200毫升）

[A]
伍斯特沙司② 1大匙
番茄酱 2大匙
盐、胡椒 各少许

1. 培根切碎。
2. 洋葱、胡萝卜、芹菜切丁，每块边长3～4毫米。
3. 蘑菇切片，切成4～5等份即可。
4. 向平底锅中倒入橄榄油，油热后，将蒜、培根下锅烹炒。放入肉馅，翻炒，撒盐和胡椒。将洋葱、胡萝卜、芹菜放入锅中，翻炒均匀，最后放入蘑菇并炒匀。
5. 向锅中倒入红葡萄酒，煮沸后加入半冰沙司和番茄汁。小火煮20～25分钟后，用A调味。

①半冰沙司：将肉和蔬菜炒熟加香料长时间熬制而成的沙司，多为褐色。
②伍斯特沙司：又称英国黑醋，起源于英国，味道酸甜微辣，为黑褐色。

Tofu lasagna

[Serves 4]

---

2 packs soft tofu
2 cups (400ml) meat sauce (see the recipe below or use store-bought)
1/4 cup (50ml) heavy cream
150g pizza cheese
2 tbsp olive oil
salt and pepper

1. Wrap the tofu in a paper towel for about 15 minutes to drain. Then, cut each piece into five slices.
2. Put the olive oil in a frying pan and turn the heat on medium. Place the tofu in the frying pan and season with salt and pepper. Brown the tofu slowly on both sides.
3. Preheat the oven to 230°C.
4. Spread a thin layer of meat sauce on the bottom of a baking dish. Place the tofu on top. Pour the cream over the layer of tofu. Repeat this process, and sprinkle the cheese on top.
5. Bake in the oven for 15-20 minutes.

Meat sauce

[Serves 4]

500g ground meat (mixed beef and pork)
50g bacon
1 onion (200g)
1/2 small carrot (50g)
1/2 stalk celery (50g)
1 pack mushrooms (100g)
1 clove garlic (finely chopped)
2 tbsp olive oil
salt, pepper
1/2 cup (100ml) red wine
1 can (290g) demi-glace sauce (store-bought)
1 cup (200ml) tomato juice

[A]
1 tbsp Worcester sauce
2 tbsp tomato ketchup
salt, pepper

1. Finely chop the bacon.
2. Cut the onion, carrot, and celery into 3-4mm squares.
3. Slice each mushroom into 4-5 pieces.
4. Heat the olive oil in a frying pan and stir-fry the garlic and bacon. Add the ground meat and stir-fry. Lightly sprinkle salt and pepper. Add the onion, carrot and celery and stir. Add the mushrooms and stir-fry.
5. Pour in the red wine and let it come to a boil. Add the demi-glace sauce and tomato juice, and let it simmer over low heat for 20-25 minutes. Season with [A].

四季豆炒肉

Stir-fried string beans and ground pork

　　这是我丈夫非常喜欢的一道菜，经常出现在我们家的餐桌上。这个食谱不仅适用于四季豆炒肉，也适用于茄子、青椒、荷兰豆炒肉。有时候我会专挑四季豆吃，导致最后剩下了很多肉馅，这种时候可以把肉馅浇在米饭上，或者放进拉面、炒饭里混着吃，非常美味。

I often cook this dish because my husband loves it. Instead of using string beans, you can also cook it with eggplant, green pepper or snow peas. Sometimes I end up with only the meat left on the plate after I have eaten up the beans. But, in that case, it is a good idea to put the meat on top of the white rice or to use it for extra flavor when you cook *ramen* noodles or fried rice.

四季豆炒肉

[用料4人份]

四季豆400克

猪肉馅150克

葱末 半根葱的量

姜末2大匙

蒜末1大匙

色拉油1大匙

绍兴酒或其他酒1大匙

酱油4大匙

红辣椒圈1～2根辣椒的量

芝麻油1大匙

1. 四季豆去筋,用水煮。之后用流水冲凉,沥干水分。斜刀切至适当大小,方便食用即可。
2. 向平底锅中倒入色拉油,油热后将葱、姜、蒜末下锅,翻炒爆香后加入猪肉馅,继续烹炒。
3. 肉馅炒熟后,向锅中倒入绍兴酒,放入四季豆,继续翻炒。加入酱油,加入红辣椒圈下锅翻炒。最后淋芝麻油提味。

Stir-fried string beans and ground pork

[Serves 4]

400g string beans
150g ground pork
1/2 finely chopped Japanese leek
2 tbsp finely chopped ginger
1 tbsp finely chopped garlic
1 tbsp vegetable oil
1 tbsp *shokoshu* (Chinese sake) or Japanese sake
4 tbsp soy sauce
1-2 red chili peppers, chopped
1 tbsp sesame oil

1. Remove the strings of the beans and boil. Chill under cold running water. Pat dry and cut diagonally into bite-sized pieces.
2. Heat the oil in a frying pan. Add the leek, ginger, and garlic, and stir-fry. When you can smell the aroma of the leek, ginger and garlic, add the ground pork and stir-fry.
3. When the pork is cooked, add the *shokoshu* and string beans, and continue to stir-fry. Add the soy sauce and red peppers and mix thoroughly. Finish by drizzling the sesame oil over it.

甜咸土豆

"Kofuki-imo"
(salty-sweet flavored potatoes)

我至今还记得第一次在英国制作这道菜肴时英国友人惊讶的表情。他们好像从未想过可以用酱油和砂糖将土豆煮得又甜又咸。但是真正尝到嘴里之后,他们又给了很高的评价。最后放进去的黄油是美味的关键。

When I first served this dish to my British friends in the U.K., I remember everyone was very surprised. This potato recipe, cooked with soy sauce and sugar, was totally unfamiliar to them. But once they ate it, they all loved it. A scoop of butter that I add before serving it is the key. It gives a rich flavor to the taste.

甜咸土豆

[用料4人份]

土豆4个（600克）
砂糖3大匙
酱油2大匙
黄油20克

1. 碗中加入砂糖和酱油，搅拌均匀至砂糖充分溶解。
2. 土豆去皮后，将每个土豆切成4块。把土豆放入锅中，加水至没过土豆为止，开火。煮沸后盖上锅盖，转小火，继续加热10～15分钟至土豆变软。
3. 将煮土豆的水全部倒出后，再次开火加热，烤干水分。将步骤1制作的酱汁倒入锅中，快速搅匀，最后放入黄油，酱汁充分渗入所有土豆后关火。

"Kofuki-imo"
(salty-sweet flavored potatoes)

[Serves 4]

4 potatoes (600g)
3 tbsp sugar
2 tbsp soy sauce
20g butter

1. Combine the sugar and soy sauce in a bowl, and mix well until the sugar dissolves.
2. Peel the potatoes and cut them into quarters. Put the potatoes in a pan, cover with water, and turn on the heat. When it comes to a boil, put a lid on, turn down the heat and simmer for 10-15 minutes in low heat until tender.
3. Drain the potatoes well, put it over the heat again and cook off the moisture. Add the sauce from step 1, and stir briskly while coating the potatoes with the sauce. Add the butter at the end and toss the potatoes, and turn off the heat.

这道菜我已经做了三十多年了，它也是我丈夫特别爱吃的菜肴之一。这道烤菜不需要制作白酱，只需要将生奶油倒在食材表面，再放入烤箱即可。注意一定要事先用微波炉加热土豆至柔软状态。这道菜的做法看似复杂，实则简单，建议大家尝试一下。

I have been cooking this gratin for over 30 years now and this is one of my husband's favorites. Instead of white sauce, I simply pour heavy cream over the ingredients and bake it in the oven. The important point is to put the potatoes in the microwave beforehand, until they are tender. It may look difficult, but in fact this dish is very easy. Please try it for yourself.

日式葱香土豆烤菜

Japanese leek and potato gratin

日式葱香土豆烤菜

[用料 1人份]

土豆 3个（400克）
葱 2根
鳀鱼 2～3片
奶酪（比萨用）150～120克
生奶油 1杯（200毫升）
盐、胡椒 各适量

1. 烤箱预热至220～230摄氏度。
2. 土豆去皮，切成圆形或半月形，每片厚度5～6毫米。土豆泡水后沥干水分。
3. 在耐热容器中铺几张厨房纸巾后，均匀地放入土豆。轻轻地用保鲜膜包住容器口，在微波炉（600瓦）中加热6～8分钟，至土豆变软。取下保鲜膜和厨房纸巾。
4. 用手撕碎鳀鱼。葱纵向切段，每段长度4～5厘米。
5. 把鳀鱼块和葱段夹在土豆中间。
6. 在生奶油中加入盐和胡椒调味。把调过味的生奶油全部倒在土豆上，不要留空隙。撒上奶酪，放入烤箱烤制20～25分钟。

Japanese leek and potato gratin

[Serves 4]

3 potatoes (400g)
2 Japanese leeks
2-3 anchovy fillets
150-200g pizza cheese
1 cup(200ml) heavy cream
salt and pepper --- to taste

1. Preheat the oven to 220-230°C.
2. Peel the potatoes and slice them into 5-6mm-thick rounds (*wagiri*) *or* half-moons (*hangetsu-giri*). Soak in water and drain well.
3. Place a few pieces of paper towel on a heat-resistant dish. Put the potatoes on it evenly. Cover with plastic wrap loosely. Microwave for 6-8 minutes until soft. Remove the plastic wrap and paper towel.
4. Tear the anchovies into small pieces. Cut the Japanese leeks into 4-5cm-long pieces, and thinly slice lengthwise.
5. Put the anchovies and the Japanese leeks between the layers of potatoes.
6. Season the heavy cream with salt and pepper and pour over the potatoes evenly. Sprinkle with cheese on top and bake in the oven for 20-25 minutes.

炸浸[1] 蔬菜

Deep-fried vegetables in *mentsuyu*

[1] 日本常见烹饪方法,将炸好的食材浸泡于高汤等调味汁中。

平时，我很喜欢制作各种各样的酱汁。方便实用的日式荞面汁就是其中之一，不仅可以将它用于面食，也能拓展应用到炖菜和盖饭里。现在我向大家介绍的这道菜，是可以提前做好、放着备用的炸浸蔬菜。除了食谱上提到的蔬菜之外，也可以根据个人喜好加入莲藕、牛蒡、薯类等食材。

I enjoy making various kinds of sauces in my kitchen, but "mentsuyu" especially comes in handy because it can be used not only for noodles but also for many recipes such as simmered dishes or "donburi" (a bowl of rice topped with meat or vegetable and sauce). In this book, I will show you how to make deep-fried vegetables in *mentsuyu* that can be kept in the refrigerator. For this recipe, you can use lotus root, burdock root, or various kinds of potatoes.

炸浸蔬菜

[用料4人份]

长茄子 3个
南瓜 1/8个
绿芦笋 4根
红彩椒 1/2个
黄彩椒 1/2个
四季豆 50克
日式蘸面汁 2杯（400毫升）
(制作方法参阅第225页)
煎炸油 适量

1. 茄子去蒂。先沿中线将茄子切成左右两半，再分别将左右两半各自切成4等份。将切好的茄子放入清水中浸泡约5分钟，捞出，擦干水分备用。
2. 南瓜去籽，切块，每块厚度约1.5厘米，大小方便食用即可。在耐热容器内铺上厨房纸巾，放入南瓜块，放入微波炉（600瓦）加热1～2分钟。
3. 绿芦笋去根，去鳞片。把芦笋斜刀切成4等份。
4. 沿中线将两种彩椒对半切开，去籽，切细丝。
5. 四季豆去筋，斜刀切成两半。
6. 将日式蘸面汁倒入平底方盘或其他浅口容器。
7. 蔬菜下锅油炸，捞出，擦干多余油分。趁热放入日式蘸面汁中浸泡。
8. 冷食热食均可。可在冰箱冷藏保存2～3天。

Deep-fried vegetables in *mentsuyu*

[Serves 4]

3 eggplants
1/8 pumpkin (*kabocha* squash)
4 spears green asparagus
1/2 red pepper
1/2 yellow pepper
50g string beans
2 cups(400ml) *mentsuyu* (see page 225)
vegetable oil
--- for deep-frying

1. Remove the stems of the eggplants and cut them in half lengthwise. Then cut them into 4 pieces. Soak the eggplants in water for about 5 minutes and pat dry.
2. Remove the seeds of the pumpkin and cut into 1.5cm-thick, bite-sized pieces. Place a paper towel on a heat-resistant plate, put the pumpkin pieces on top, and microwave at 600W for 1-2 minutes.
3. Cut off the woody part of the asparagus and remove the scales. Cut each asparagus into 4 pieces diagonally.
4. Cut each pepper in half lengthwise and remove the seeds. Cut in half crosswise and slice into thin strips.
5. String the beans, and cut them diagonally in half.
6. Put the *mentsuyu* in a shallow dish.
7. Deep-fry all the vegetables in oil and drain them. Soak them in *mentsuyu* while still hot.
8. Serve hot or cold. You can keep the vegetables in the refrigerator for 2-3 days.

便当 | Bento (box lunch)

便 当

在小小的盒子中装满美味的米饭和多彩的配菜,这就是日式便当。

打开盖子之前的"小激动"也是日常生活中的"小确幸"。

"bento"这个英语词汇的诞生也说明了便当在世界范围内广受欢迎。

我做了将近14年的便当。

从孩子上幼儿园到高中毕业。

现在回过头来看,会不由得感叹自己在制作便当的过程中学到了很多,受益匪浅。

"如何装盘会让食物看上去更有食欲,会让孩子吃得更开心?"

这样想着的同时,

我每天也在颜色搭配、口味平衡、装盘美感等方面下了很多功夫。

为了能够不慌不忙地有效利用每天早上忙碌且紧张的时间,

我发现可以巧用家中的常备菜,

也可以在前一天做晚饭的时候顺便准备第二天早上需要的食材。

在认真准备孩子的便当的同时,我的厨艺也不知不觉地进步了不少。

制作便当没有特定的规则。

所以,如果本书介绍的食谱中刚好有您喜欢的菜品的话,

不妨试着将它放进便当盒里,把这当成厨艺的起点。

我认为,将菜品装进便当盒之后,能够体会到一种别样的美味。

Bento (box lunch)

The Japanese bento, or box lunches are tightly packed
with colorful foods and rice.
It has fascinated many people overseas, who find it fun
and exciting to open the bento box, and the name "bento"
has become widely acknowledged throughout the world.
I started making bento for my children
when they entered kindergarten and continued
for 14 years till they graduated from high school.
Looking back, I realize that I learned a lot through bento-making.
"What's the best way to pack it?"
"How can I make one that they would love?"
I was constantly thinking these thoughts,
and was trying different methods on a daily basis to make a bento
that had a good balance in taste and color,
and looked appealing to the eye.
Because mornings are always busy, I have developed time-saving
and efficient techniques such as packing into the bento box side dishes
that are usually kept in the fridge for days,
or doing the preparation for bento
while making dinner the night before.
I realized that all these years of putting a lot of effort into
making bento has made me better at cooking.
There are no ground rules in bento-making.
If you find a recipe you like in this book,
please start by putting that into the bento.
You will find that it will have a different appeal
once packed in a bento.

日式烹调器具
Japanese cooking utensils

［卷帘］
Makisu

［寿司饭台、木勺］
Handai Shamoji

［擦菜板］
Oroshigane

［芝麻煎锅］
Goma-iri

[长筷、装盘筷]
Saibashi Moritsuke-bashi

[土锅]
Donabe

[蒸食器]
Mushiki

日式烹调器具

Japanese cooking utensils

制作日式菜肴的时候会用到各种各样的烹调器具。它们无一不是由日本手工技师们小心仔细地制作出来的。我经常觉得这些工具越用越顺手，也越发认为应该带着对手工技师们的感激之情，珍惜地使用这些工具。

There are many different kinds of utensils for Japanese cooking. Each is made with great attention to detail by the hands of Japanese artisans, and the more I use them, the more I am impressed with how practical they are. Every time I use the tools, I feel a sense of appreciation towards all the craftsmen throughout Japan.

[寿司饭台、木勺]

Handai (wooden sushi tub, wide cooking bowl)
Shamoji (rice paddle)

制作寿司醋饭的工具。木制饭台能够吸收多余的水分，从而保证寿司醋饭的口感。

Tools used for making sushi rice. They absorb the excess water and help make delicious sushi rice.

[卷帘] *Makisu* (rolling mat)

主要用来卷制寿司，通常为竹子材质。也可用来调整日式甜蛋卷的形状。在需要挤干过水蔬菜中多余水分的时候，用卷帘包住蔬菜能够起到一定的保护效果。

Mainly used for making sushi rolls. Made of bamboo. It can also be used to arrange the shape of the *tamago-yaki* (rolled Japanese omelet), squeeze out excess water from boiled vegetables, etc.

[擦菜板] *Oroshigane* (grater)

日本的擦菜板可以称得上是种类丰富。图片中左侧的工具叫作"鬼竹"，是磨白萝卜的常用工具。经它磨制的白萝卜泥不够细腻，但也正因如此，擦菜过程中食材基本不会流失水分，所以用它磨出的白萝卜泥口感松软，十分美味。图片中央的铜制工具是由工匠手工制作的。图片中右侧的工具专门用来磨芥末泥。经它磨制的芥末泥充满空气感，口感十分顺滑。

There are various types of graters in Japan. The one on the left is called *Oni-oroshi*, and it is used for coarsely grating daikon. By coarsely grating, loss of moisture is kept at a minimum, making for a fluffy, tasty daikon-oroshi. The one in the center is made of copper, crafted by the hands of an artisan. The one on the right is made from shark skin, and is exclusively used for grating wasabi. By using this, you can make airy and creamy grated wasabi.

[芝麻煎锅]

Goma-iri (sesame seed toaster)

煎炒芝麻的专用厨具。虽说也可以用普通的平底锅煎炒芝麻，但是用这种专用工具的话，不仅芝麻不容易焖，而且成品的香味也会更加浓。

A utensil specifically for toasting sesame seeds. You can substitute it with a frying pan, but this tool will allow you to toast sesame seeds without burning them, and will enhance the fragrance.

[长筷、装盘筷]

Saibashi Moritsuke-bashi (cooking chopsticks and serving chopsticks)

这两双筷子可不是用来吃饭的，它们是专门用来烹饪的厨具。长筷可以用于炒菜等所有烹饪场合。装盘筷则是专门用来装盘的筷子，筷子前端很细，能够夹起各种细小食材。筷子尾端有一定斜度，可以用这一侧夹起比较柔软的食材，并且不会将食材弄碎。另外，也可以用装盘筷的尾端代替刮刀使用。

These chopsticks are not for eating a meal, but for cooking or serving a meal. *Saibashi* is used for stir-frying and for cooking in general. *Moritsuke-bashi* is used exclusively for serving. The narrow end has a sharp, pointy edge, enabling it to pick up even the smallest items. The opposite end is slanted, so it can be used like a spatula, or to pick up soft items without squashing.

[土锅] *Donabe* (Japanese clay pot)

其实就是陶制砂锅。常用于制作火锅。由于土锅保温性能好，我也经常用它煮饭、煲粥、炖菜。

This is a ceramic pot. It is often used for making hot pot dishes but because of its excellent heat retention ability, it is quite versatile and I use it for cooking rice and simmering dishes, and so on.

[蒸食器] *Mushiki* (steamer)

日式和食中蒸制的菜肴很多，蒸食器是不可或缺的厨具之一。茶碗蒸在国外格外受欢迎。用蒸食器蒸制的菜肴口感温和，蔬菜也会变得更加鲜美。

Many Japanese dishes involve steaming in the process, so the steamer is an important cooking tool. *Chawan-mushi* savory custard is especially popular among people overseas. The steamer enhances the flavor of vegetables, and gives dishes a mild taste.

研 磨 钵
Mortar

　　研磨钵这种东西型号不一，有大有小。因为母亲以前经常制作会用到芝麻的菜品，所以我自幼时起便经常帮她研磨芝麻。照片上的这个是母亲以前经常使用的研磨钵，这个研磨钵用得年头太久，导致上面有一些磨损和开裂的地方，但是即便如此，我也想要继续使用它。

　　市面上也可以买到研磨好的碎芝麻，但其实一旦自己磨过一次芝麻，就会惊讶于手磨芝麻的美味。今后我也会珍惜地使用这个研磨钵，希望以后能将它传给孩子们。如果将来孩子们用它研磨食材的时候，能够睹物思人想起我，那我一定会很开心。

Mortars vary in size, from small to large. My mother used to make dishes with sesame seeds quite often. So to help her out, I have been grinding sesame seeds in a mortar since I was little. This mortar is the one my mother used. It has been in use for such a long time that is has cracks in it, but I intend to continue using this mortar, passed down from my mother. There are ready-made sesame pastes available in stores, but nothing compares to the one you make with your own hands using a mortar. I intend to use this mortar with care, and pass it on to my children. I hope it will remind them of me when they use it someday.

日式牛肉饭

Gyudon
(beef on rice)

牛肉饭是日本极受欢迎的快餐之一。主要食材非常简单，只需洋葱和牛肉即可，即使身处海外也能轻易地采买食材，进行制作。用白葡萄酒炖过的牛肉会变得很软，因此，哪怕选用牛身上肉质较硬的部分来进行烹饪，也照样能做得很好吃。

Gyudon is one of the most popular fast foods in Japan. The main ingredients are only onion and beef, so making this outside of Japan is easy. Even inexpensive, tough beef can be delicious because simmering with white wine softens the beef.

日式牛肉饭

[用料 4人份]

牛肉薄切片 500克
洋葱 4个（800克）
白葡萄酒 2杯（400毫升）
水 1杯（200毫升）
酱油 3/4杯（150毫升）
甜料酒 1/2杯（100毫升）
砂糖 4人匙
米饭 适量
日式红姜 适量

1. 洋葱切成条，每条宽度约为1厘米。
2. 牛肉片切段，每段长度为6~7厘米。
3. 锅中倒入水和白葡萄酒，搅匀，开中火，煮沸后放入牛肉。撇去浮沫，小火炖10～15分钟。若肉质较硬，则适当延长煮炖时间。
4. 锅中倒入酱油、甜料酒、砂糖后，盖上落盖，继续炖10分钟左右。
5. 洋葱下锅，待洋葱变透明后关火。静置入味。
6. 连汤带肉倒在热乎乎的米饭上，可搭配日式红姜食用。

Gyudon (beef on rice)

[Serves 4]

500g thinly sliced beef
4 onions (800g)
2 cups (400ml) white wine
1 cup (200ml) water
3/4 cup (150ml) soy sauce
1/2 cup (100ml) mirin
4 tbsp sugar
cooked rice
red pickled ginger to taste

1. Slice the onions into 1cm-thick *hangetsu-giri* pieces (see page 280).
2. Cut the beef into 6-7cm lengths.
3. Heat the water and white wine on medium heat. When it comes to a boil, add the beef. Skim the surface of the broth. Turn the heat to low and simmer it for 10-15 minutes. If the beef is still hard, simmer it a little longer.
4. Add the soy sauce, mirin, and sugar, and cover it with a drop-lid. Simmer for about 10 minutes.
5. Add the onions, and simmer until translucent. Turn off the heat and let it stand.
6. Serve this along with the sauce on top of hot cooked rice. Garnish with a little red pickled ginger.

亲子盖饭 [1]

Chicken and egg on rice

[1]亲子盖饭:又称滑蛋鸡肉饭,因以鸡肉和鸡蛋为主要原料而得名。

　　大家是否有过每隔一段时间就特别想吃某道菜品的经历？对我来说，亲子盖饭就是这样的存在。我平时会将鸡肉冷冻备用，这样一来，突然特别想吃亲子盖饭的时候就很方便。对于这道菜来说，鸡蛋顺滑柔软的口感至关重要，因此，鸡蛋下锅后一定即刻关火并盖上锅盖，通过余热将鸡蛋焖至半熟状态。

Have you ever experienced a sudden intense craving for a certain dish? For me, *oyako-don* is one such dish. This recipe is easy. So, whenever I feel like eating *oyako-don*, I can prepare it right away because I always have some chicken in my freezer.

To keep the smooth and soft texture of egg, which is very important, turn off the heat after adding the egg and put a lid on the pan to half-cook the egg with residual heat.

亲子盖饭

[用料2人份]

鸡腿肉（带皮）1小块（200克）
洋葱 1/2个
鸡蛋 4个
米饭 适量
鸭儿芹（切碎）适量

[混合调味汁]
日式高汤 1/2杯（100毫升）
酱油 3大匙
砂糖 2大匙
甜料酒 2大匙

海苔、腌菜 各适量

1. 洋葱切细条，每条宽度为6～7毫米。鸡肉切成一口大小，偏小为宜。
2. 制作混合调味汁，将所需的调味料搅匀。
3. 分两次打散鸡蛋，每次使用2个鸡蛋，蛋液放入不同的碗中备用。
   *接下来的步骤为单人份亲子盖饭的制作方法，本食谱食材适用于2人份量，分两次重复以下步骤即可。
4. 将一半的混合调味汁倒入锅中，煮开。放入一半的鸡肉继续煮炖稍许后，再放入一半的洋葱煮1～2分钟。
5. 趁锅中食材再次煮开的时候，倒入2/3碗蛋液。盖上锅盖，待蛋液凝固至半熟状态后，从锅的边缘倒入剩余的1/3碗蛋液。关火，放入适量鸭儿芹，盖上锅盖，焖熟。（焖至鸡蛋半熟状态最佳）
6. 连汤汁一起浇到米饭上，并用海苔和腌菜加以点缀。

Chicken and egg on rice

[Serves 2]

1 boneless small chicken thigh with skin(200g)
1/2 onion
4 eggs
cooked rice
mitsuba (trefoil), chopped
--- for garnish

[dashi mixture]
1/2 cup (100ml) dashi
3 tbsp soy sauce
2 tbsp sugar
2 tbsp mirin

nori seaweed, pickles
--- for garnish

1. Slice the onion into 6-7mm-thick pieces. Cut the chicken into smaller bite-sized pieces.
2. Combine the ingredients for the dashi mixture.
3. Beat 2 eggs in 2 separate bowls.
4. Put half of the dashi mixture into a pan and bring to a boil. Add half of the chicken and cook for a short time. Add half of the onion slices and simmer for 1-2 minutes.
5. While it is simmering, pour two-thirds of the 2 beaten eggs and put on a lid. When it is half-cooked, pour the remaining one-third of the beaten eggs evenly around the rim of the pan. Turn off the heat, add the *mitsuba*, and allow to settle with the lid on. (The egg should be slightly runny.) Repeat the process to make another serving.
6. Divide the rice in individual bowls and put the chicken and egg on top. Garnish with some nori and pickles.

生 姜 饭

Ginger rice

　　不是把生姜掺在米饭里一起煮，而是在煮好的米饭中加入生姜。这样一来，生姜的香味更加浓郁。我经常用生姜饭捏饭团，搭配上香香脆脆的海苔，十分美味。

The ginger is added after the rice is cooked, not cooked with it. By doing this, the aroma of the ginger is stronger. I often eat this dish with crisp nori seaweed.

生姜饭

[用料 4人份]

大米 2 杯（400毫升）

生姜 30 克

日式油豆腐皮① 2 片

淡口酱油 2 大匙

甜料酒 1 大匙

酒 1 大匙

日式高汤 适量

盐 少许

1. 大米洗净沥干，静置 10～15 分钟。
2. 生姜切末。
3. 将日式油豆腐皮放入沸水中汆一下，捞出后轻轻地挤干水分。将日式油豆腐皮切小丁，边长约为 5 毫米。
4. 将淡口酱油、甜料酒、酒倒入容器混合后，加入适量日式高汤，制成总量 2 杯（400毫升）的调味汁。
5. 选用材质较厚的锅具，向锅中加入大米后，将油豆腐皮置于其上，加入步骤 4 调制的调味汁。
6. 盖上锅盖，开火，煮开后转小火煮 10～12 分钟。关火，焖 10 分钟左右。加入生姜，简单地搅拌几下。可根据个人口味加少量食盐调味。

---

①日式油豆腐皮：将豆腐切成薄片后油炸或油煎制成的食品。

Ginger rice

[Serves 4]

2 cups (400ml) rice
30g ginger
2 pieces *abura-age*
(thinly sliced deep-fried tofu)
2 tbsp light soy sauce
1 tbsp mirin
1 tbsp sake
dashi
salt

1. Wash the rice well and drain. Let it stand for 10-15 minutes.
2. Mince the ginger finely.
3. Pour hot water over the *abura-age* to remove excess oil, squeeze lightly, and pat dry. Cut it into 5mm square pieces.
4. Combine the light soy sauce, mirin, and sake, and add enough dashi to make 400ml of sauce.
5. Put the rice in a heavy pan. Add the *abura-age* and dashi mixture.
6. Cover and place over high heat. When it comes to a boil, turn down the heat and cook for 10-12 minutes. Turn off the heat, and let it stand for about 10 minutes to allow settling. Then mix in the ginger. Season to taste with a little salt if necessary.

猪肉蔬菜什锦饭

Steamed rice with pork and vegetables

虽说我平时也很喜欢味道温和的什锦饭，但这道口味偏重的什锦饭也是我的心头好。提前腌制猪肉，鲜美的肉汁和香喷喷的米饭融为一体，一锅香气扑鼻的什锦饭就煮好了。另外，这道什锦饭里面既有猪肉又有黄豆，可以说是份量满满。

Sometimes I like a light taste for *takikomi gohan* (rice steamed with vegetables, meat, or fish). However, the one I introduce here has a rich flavor. The meat should be deeply dressed with seasoning beforehand so that its "umami" (soup from the seasoned meat) can soak well into the rice during the cooking process. As it uses soybeans as well as meat, this recipe is hearty.

猪肉蔬菜什锦饭

[用料 4 人份]

大米 2 杯（400 毫升）
猪肉薄切片 150 克

[**猪肉腌料**]
酱油 2 大匙
砂糖 1/2 大匙
生姜汁 1 小匙

胡萝卜 1/2 根
鲜香菇 2～3 个
水煮黄豆 1 杯（200 毫升）
酱油 1 大匙
甜料酒 1 大匙
酒 1 大匙
日式高汤 适量
盐 少许

1. 大米洗净沥干，静置 10～15 分钟。
2. 猪肉片切成小丁，用酱油、砂糖、生姜汁腌制。
3. 胡萝卜去皮，切成银杏叶形，偏小为佳。香菇去蒂，切块，每块边长约 2 厘米。
4. 锅中放米，分散地放入胡萝卜、香菇、水煮黄豆，最后在顶部铺上猪肉丁，注意猪肉丁平铺一层即可。
5. 把酱油、甜料酒、酒倒入量杯，再倒入适量日式高汤至总量 400 毫升。加入食盐调味，从锅边轻轻地倒入。盖上锅盖，大火加热。煮开后转小火，焖 10～12 分钟。
6. 关火，静置约 10 分钟。打开锅盖。搅拌均匀。

Steamed rice with pork and vegetables

[Serves 4]

2 cups (400ml) rice
150g sliced pork

[marinade sauce for pork]
2 tbsp soy sauce
1/2 tbsp sugar
1 tsp juice from grated ginger

1/2 carrot
2-3 fresh shiitake mushrooms
1 cup (200ml) boiled soybeans
1 tbsp soy sauce
1 tbsp mirin
1 tbsp sake
dashi
salt

1. Wash the rice and drain in a strainer. Let it stand for 10-15 minutes.
2. Chop the pork into tiny pieces and marinate in soy sauce, sugar, and juice from grated ginger.
3. Peel the carrot and cut into small *icho-giri* pieces (see page 280). Cut off the stems of the shiitake mushrooms and cut into 2cm squares.
4. Put the rice in a pan and scatter the carrot, shiitake mushrooms, and soybeans. Top it with the marinated pork, making sure they don't overlap.
5. Pour the soy sauce, mirin, and sake into a measuring cup. Add the dashi until it measures 400ml and season with salt. Gently pour the dashi mixture onto the rice from the rim of the pan. Put a lid on and turn the heat on high. When it comes to a boil, turn the heat down to low and cook for 10-12 minutes.
6. When the rice is cooked, turn off the heat and let it stand for about 10 minutes. Stir the rice well.

三色盖饭

Three-color rice bowl

这是一道在日本家庭中常见的典型日式家常盖饭。接下来要介绍的做法是我跟母亲学习的。先将肉馅下锅烹煮，再用煮肉馅的汤汁煮饭。

这样虽然比较花时间，但也确实能让米饭变得更美味。我很喜欢这个做法，平时也都是这么操作的。注意一定要仔细炒制鸡蛋，并在米饭上放很多荷兰豆，记得要把荷兰豆切得细一点。

This is one of the Japanese home-cooked dishes which is served in typical Japanese families. Here, I introduce the recipe that my mother taught me. After quickly simmering ground meat in a pan, use the original broth from boiling the meat to cook the rice.
This takes time and effort, but it adds richness to the taste and I like it this way. I believe the important point here is to scramble the eggs finely and to use plenty of thinly sliced snow peas on top of white rice.

三色盖饭

[用料 4人份]

大米 2杯（400毫升）
（洗净沥干，参阅第014页）
鸡肉馅 300克
荷兰豆 100克
日式高汤 适量
盐 少许
海苔 适量
日式红姜 适量

[A]
酱油 1大匙
酒 1大匙
甜料酒 1大匙

[B]
酱油 2+1/2～3大匙
酒 1大匙
甜料酒 2大匙
砂糖 1+1/2～2大匙

[炒蛋]
鸡蛋 4个
砂糖 1+1/2～2大匙
酒 2大匙
盐 少许

1. 锅中倒入1杯（200毫升）日式高汤和[A]的调料，煮开后加入鸡肉馅继续煮制。鸡肉煮熟后，过滤，分离汤汁和肉馅。

2. 将步骤1的汤汁倒入量杯，再倒入适量日式高汤至总量400毫升。加少许食盐调味。

3. 用电饭煲煮饭。锅中加米，倒入步骤2调制的混合汤汁，煮饭。
   *如果需要用锅具煮饭，请参阅第14页的煮饭方法。

4. 另起一锅，倒入[B]的调料，煮开后倒入步骤1的鸡肉馅，煮至汤汁基本收干。

5. 制作炒蛋。将鸡蛋于碗中打散，加入砂糖、酒、盐并搅匀。倒入锅中，以中火加热。鸡蛋边缘凝固后转小火，用4根长筷搅拌至鸡蛋炒熟。

6. 锅中加水烧开，荷兰豆迅速焯水后捞出，放进冷水。沥干水分后，斜刀切成细丝。

7. 碗中盛饭，将鸡肉馅、炒鸡蛋、荷兰豆细丝放在米饭上。撒海苔碎，旁边摆放日式红姜。

Three-color rice bowl

[Serves 4]

2 cups (400ml) rice, washed and drained (see page 015)
300g ground chicken
100g snow peas
dashi
salt
nori seaweed
--- for garnish
red pickled ginger
--- for garnish

[A]
1 tbsp soy sauce
1 tbsp sake
1 tbsp mirin

[B]
2+1/2-3 tbsp soy sauce
1 tbsp sake
2 tbsp mirin
1+1/2-2 tbsp sugar

[scrambled eggs]
4 eggs
1+1/2-2 tbsp sugar
2 tbsp sake
salt --- a little

1. Combine 1 cup/200ml dashi and the ingredients of [A] in a pan and bring to a boil. Add the ground chicken and simmer. When the chicken is cooked, strain it to separate the soup and chicken.
2. Put the soup into a measuring cup and add more dashi until it measures 400ml. Season with a little salt.
3. Put the rice in a rice cooker and add the dashi mixture. Turn on the rice cooker.
   * If you want to use a pan instead of a rice cooker, see page 015.
4. Combine the ingredients of [B] in another pan and bring to a boil. Add the cooked chicken and simmer until the sauce is almost gone.
5. Make scrambled eggs : Beat the eggs in a bowl. Add the sugar, sake, and salt, and mix together. Pour the egg in a pan and turn the heat on medium. When the rim of the egg has just set, turn the heat down to low and stir the mixture with 4 chopsticks until it is cooked.
6. Blanch the snow peas and chill in cold water. Drain and cut diagonally into thin strips.
7. Divide the rice into individual bowls. Arrange the chicken, eggs, and snow peas on top. Garnish with some nori and red pickled ginger.

炸猪排盖饭
"Katsu-don"
(pork cutlet on rice)

　我喜欢鸡蛋半熟的炸猪排盖饭。图上这个可单手使用的小锅比我的年纪都大，是我家的盖饭专用锅，炸猪排盖饭和亲子盖饭都会用到它。用这个小锅刚好可以做一人份的盖饭，直接将配料倒在米饭上，非常方便。

I like my eggs half-cooked for *katsu-don*. This small skillet with a handle is made specifically for cooking *katsu-don* or *oyako-don*. My mother used this before I was born, and I use it to this day. It's a good size for cooking just enough for one *donburi* bowl, and is designed to make it easy to directly pour the contents of the pan over rice.

炸猪排盖饭

[用料1人份]

猪排 1块
洋葱 1/4个
鸡蛋 2个
日式高汤 1/2杯（200毫升）
砂糖 1大匙
酱油 2大匙
甜料酒 1大匙
米饭 适量

1. 将猪排切至适当大小，方便食用即可。
2. 洋葱切薄片。
3. 鸡蛋打散、搅匀。
4. 把日式高汤、砂糖、酱油、甜料酒倒入浅锅，中火加热。先放洋葱，煮制稍许后放入猪排，水开后倒入蛋液，盖上锅盖，加热约1分钟后关火。
5. 将4浇在米饭上即可。

"Katsu-don" (pork cutlet on rice)

[Serves 1]

1 *tonkatsu*
1/4 onion
2 eggs
1/2 cup(100ml) dashi
1 tbsp sugar
2 tbsp soy sauce
1 tbsp mirin
cooked rice

1. Cut the cutlet into bite-sized pieces.
2. Slice the onion thinly.
3. Beat the eggs.
4. Combine the dashi, sugar, soy sauce, and mirin in a shallow pan. Warm it on medium heat and add the onion. Simmer for a short time, add the cutlet, and continue cooking for a while. When it comes to a boil, pour the beaten egg over the cutlet and cover with a lid for 1 minute before turning the heat off.
5. Put the rice in a serving bowl and place the cooked cutlet on top.

微波炉红豆饭

Microwaved "sekihan" (azuki beans and rice)

红豆饭是日本典型的庆祝餐食，遇见需要庆祝的好事的时候，人们一般都会吃上一碗红豆饭。我们夫妻二人都非常喜欢吃红豆饭，所以它也是我家餐桌上的常客。传统做法的红豆饭需要蒸制而成。为了能够随时满足口腹之欲，我想出了这个用微波炉做红豆饭的便捷食谱。

Sekihan is usually served on celebratory occasions in Japan. But I often cook it regardless of occasion as my husband and I both love it. Traditionally, it is cooked in a steamer, but I came up with this idea to use a microwave to cook it easily whenever I feel like eating *sekihan*.

*微波炉型号不同可能会导致煮饭口感稍有不同。建议在本食谱的基础上，根据实际情况，自行调整加热时间。

*The *sekihan* might become a little bit too soft depending on the microwave type. Please adjust the cooking time as needed by using the times given here as a reference.

微波炉红豆饭

[用料 4 人份]

糯米 2 杯（400 毫升）
红豆 60 克
即食黑芝麻 适量
盐 适量

1. 红豆洗净，用充足的清水提前浸泡 2～3 小时。
2. 糯米洗净，浸泡 30 分钟左右，捞出，沥干水分。
3. 将红豆和 300 毫升清水倒入锅中，小火加热。红豆煮至偏硬。
   *如果可以用手指捏碎的话，那就是煮过头了。注意红豆不要煮太软，偏硬为佳。
4. 将红豆置于滤筛上，沥干水分。不要扔掉煮红豆的水。在煮红豆的水中加入清水，总量 300 毫升即可，放凉备用。
5. 在耐热容器中加入红豆、糯米、步骤 4 的红豆水，用保鲜膜轻轻地包住容器口。放入微波炉（600 瓦）加热约 9 分钟。
6. 取出容器，将糯米等搅拌均匀。再次包上保鲜膜，加热 2～3 分钟。
7. 盛入碗中，撒适量黑芝麻和盐。

Microwaved "sekihan" (azuki beans and rice)

[Serves 4]

2 cups(400ml) *mochigome* (glutinous rice)

60g dry azuki beans

roasted black sesame seeds --- to serve

salt --- to serve

1. Soak the azuki beans in plenty of water for 2-3 hours.
2. Wash the glutinous rice. Soak in water for about 30 minutes and drain.
3. Put the azuki beans in a pan and add 300ml water. Place the pan over low heat and simmer until the beans are barely cooked.
  * If the beans can be crushed with fingers, they are overcooked. Make sure some firmness remains.
4. Drain the azuki beans in a strainer, saving the water they were cooked in. Add more water until it measures 300ml. Set aside and cool.
5. Put the azuki beans, rice, and the water in a heat-resistant bowl. Cover it loosely with plastic wrap and microwave on 600W for about 9 minutes.
6. Remove from the microwave and stir well. Cover the bowl with plastic wrap once again and microwave for 2-3 more minutes.
7. Serve in a serving bowl and sprinkle with sesame seeds and salt to taste.

每年我都会制作各种各样的寿司料理。但是，到了女儿节①的时节，果然必不可少的还是散寿司。即使没有备齐食谱中的配料也没关系，但是请一定要试着学会制作日式蛋丝。将鸡蛋切成细丝后，只要拨散蛋丝，增加蛋丝的空气感，就可以提高整体完成度。刚开始尝试时鸡蛋煎得厚也很正常，勤加练习就可以做得很好。

I cook various kinds of sushi throughout the year. But in the *Hinamatsuri* (Doll Festival) season in early March, I never fail to make "chirashi-zushi." If you don't have all the ingredients I introduced here, don't worry about it. But please learn how to make *kinshitamago* (shredded egg crepes). Thinly shred the egg crepes, loosen the shreds, and put them on top of the sushi in a fluffy manner. This makes the dish look very gorgeous. Don't worry if at first you cannot make the egg crepes thinly. With practice you will get the knack.

---

①女儿节：每年公历3月3日，亦称桃花节、雏祭等，日本五节之一。有女孩子的人家搭设坛架，陈列人偶，供摆菱形年糕、米酒、桃花等，为家中的女孩子祈求健康、幸福。

散寿司
"Chirashi-zushi"

散 寿 司

[用料4人份]

[寿司醋饭]

大米 2杯（400毫升）

水 2杯（400毫升）

[寿司醋]

醋 1/2杯（100毫升）

砂糖 1+1/2～2大匙

盐 1小匙

[甜煮干香菇]

干香菇 8个

日式高汤 1/2杯（100毫升）

砂糖 2大匙

酱油 1+1/2大匙

酒 1大匙

甜料酒 1大匙

[醋藕]

莲藕 1节（200克）

醋 5大匙

砂糖 2大匙

盐 少许

1. 制作寿司醋。碗中倒入醋、砂糖、盐，仔细搅拌均匀。
2. 制作寿司醋饭。大米洗净，捞出，在滤筛上放置约15分钟沥干水分。将大米放入锅中，按1:1的比例倒入清水，开火煮饭。向刚出锅的米饭中倒入寿司醋，搅拌均匀后放凉备用。
3. 制作甜煮干香菇。用少量清水泡发干香菇。轻轻地挤干水分后，切掉香菇根。锅中倒入日式高汤和其他调味料，混合均匀，汤汁煮沸后放入香菇，盖上落盖，小火加热10～15分钟，煮至汤汁收少即可关火，静置入味。自然放凉后将每个香菇切成4等份。
4. 制作醋藕。莲藕去皮，切成银杏叶形，每片厚度约1厘米。将莲藕片在清水中浸泡稍许，捞出，仔细擦干水分。锅中倒入所需调料，煮沸后放入莲藕片，中火煮1～2分钟后关火，注意煮莲藕片的时候需要快速搅拌。放凉备用。

"Chirashi-zushi"

[Serves 4]

[sushi rice]
2 cups(400ml) rice
2 cups(400ml) water

[sushi vinegar]
1/2 cup (100ml) vinegar
1+1/2～2 tbsp sugar
1 tsp salt

[sweet simmered shiitake muhrooms]
8 dried shiitake mushrooms
1/2 cup (100ml) dashi
2 tbsp sugar
1+1/2 tbsp soy sauce
1 tbsp sake
1 tbsp mirin

[vinegared lotus root]
1 lotus root (200g)
5 tbsp vinegar
2 tbsp sugar
salt

1. Make sushi vinegar: Combine the vinegar, sugar, salt in a bowl, mix well and dissolve.
2. Make sushi rice: Wash the rice and drain it in a strainer. Let it stand for about 15 minutes. Put the rice in a pan, add the same amount of water to the pan as rice and cook. Pour sushi vinegar over freshly cooked rice and fold it in while cooling the rice.
3. Make sweet simmered shiitake mushrooms: Soak the dried shiitake mushrooms in water until they become soft. Lightly squeeze them to drain and cut off the stems. Combine the dashi with other ingredients in a pot and bring to a boil. Add the shiitake mushrooms, cover with a drop-lid and simmer over low heat for 10-15 minutes until the liquid is reduced. Turn off the heat and let it stand for a while to let the mushrooms absorb the sauce. When cooled, cut into quarters.
4. Make vinegared lotus root: Peel and cut the lotus root into 1cm-thick *icho-giri* (quarter-rounds) pieces. Soak in water and drain well. Combine the seasonings in a pan and bring to a boil. Add the lotus root and simmer for 1-2 minutes over middle heat while stirring quickly. Let it cool.

散 寿 司

[用料4人份]

[日式蛋丝]

鸡蛋 2个
砂糖 1大匙
酒 1/2大匙
盐 少许

色拉油 适量

[生鱼片]

金枪鱼（红肉）1块
白肉鱼（鲷鱼等）1块
水煮章鱼脚 1根

酸橘、甜醋腌生姜、海苔碎 各适量
酱油、绿芥末泥 各适量

5. 制作日式蛋丝。鸡蛋打散，加入所需调料搅拌均匀后，使用滤网等过滤蛋液。向小号平底锅中倒入色拉油，加热，用厨房纸巾将色拉油均匀地涂抹在锅底。倒入薄薄的一层蛋液，在鸡蛋变色前翻面，分两面煎制蛋皮。按相同步骤煎制多张蛋皮，至用尽全部蛋液为止。将煎好的蛋皮叠在一起，轻轻卷起，切细丝。搅散蛋丝，空气进入蛋丝后口感更加松软。

6. 鱼肉全部切片，每块边长约2厘米。

7. 将寿司醋饭盛入容器，放上大量的日式蛋丝后，再将生鱼片、甜煮干香菇、醋藕分散地放在上面。可根据个人口味放适量的酸橘片和甜醋腌生姜。在最上面撒上海苔碎。生鱼片蘸酱油和绿芥末食用。

"Chirashi-zushi"

[Serves 4]

[*kinshi-tamago*
(shredded egg crepe)]
2 eggs
1 tbsp sugar
1/2 tbsp sake
salt

vegetable oil

[sashimi]
1 *saku* block
  tuna (lean)
1 *saku* block
  white-flesh fish
  (sea bream etc.)
1 leg boiled octopus

sudachi, pickled
  ginger, crumbled
  nori seaweed
  --- for garnish
soy sauce,
  grated wasabi
  --- for sashimi

5. Make *kinshi-tamago* (shredded egg crepes): Beat the eggs in a bowl, add the sugar, sake and salt, mix well and strain. Using a paper towel, grease a small frying pan with some vegetable oil. Pour a small amount of egg mixture in the pan, and spread it out to make a thin layer. Before the surface of the egg turns color, flip it over and cook the other side as well. Remove the egg crepe from the pan. Repeat the same processs with the rest of the egg mixture. Stack up the egg crepes, roll them together and cut them into thin strips. Loosen them gently by pulling them apart to let some air in.

6. Cut the sashimi into 1.5-2cm cubes.

7. Serve the sushi rice onto a plate, top it with plenty of *kinshi-tamago*. Scatter diced sashimi, sweet simmered shiitake mushrooms, and vinegared lotus root on top. Garnish with sudachi, pickled ginger to taste. Top it with crumbled nori seaweed. Serve with wasabi and soy sauce for the sashimi.

反卷寿司

"Uramaki-zushi"
(inside-out sushi rolls)

这是我在外国的时候最常做的一道料理。按牛油果、青紫苏、蛋黄酱、蟹棒的顺序放置食材，便能卷出形状整齐、色彩漂亮的反卷寿司。

基本上可以说，很少有买不到这些食材的地方，也很少有不喜欢这个口味的人。

I often make this dish when I am overseas. The trick is to be sure to add the filling in the correct order: first the avocado, then the shiso leaves, mayonnaise, and crab sticks. This way you will make an excellent roll and the color will be beautiful. These fillings are available anywhere, and are everyone's favorites.

反 卷 寿 司

[用料 6根寿司卷]

[寿司醋饭]
米 2杯（400毫升）
水 2杯（400毫升）
米醋 1/2杯（100毫升）
砂糖 1+1/2～2大匙
盐 1小匙

蟹棒 18根
牛油果 1～1+1/2个
青紫苏 9片
烤海苔 3张
蛋黄酱 适量
即食白芝麻 适量

1. 制作寿司醋饭，具体操作请参阅第194页。
2. 海苔对半切开。将烘焙纸切至比海苔大一圈的大小。
3. 牛油果去皮去核，切梳形块，每个牛油果切成12等份。青紫苏一切为二。
4. 在卷帘上依次铺平烘焙纸、海苔和一层薄薄的寿司醋饭后，将米饭和海苔整体翻面，使海苔面朝上。
5. 在海苔中央靠近自己的位置摆放牛油果，码成一列。铺上青紫苏，涂1小匙蛋黄酱，放上蟹棒。
6. 卷寿司。卷一圈后轻轻按压定型，确保不会卷到烘焙纸后，继续卷到另一端即可。寿司卷接缝处朝下，再次轻轻按压。取下卷帘和烘焙纸。
7. 寿司表面涂满芝麻，切成6等份。

"Uramaki-zushi" (inside-out sushi rolls)

[Makes 6 rolls]

[sushi rice]
2 cups (400ml) rice
2 cups (400ml) water
1/2 cup (100ml) rice vinegar
1+1/2-2 tbsp sugar
1 tsp salt

18 crab sticks
1-1+1/2 avocado
9 shiso leaves
 (green perilla)
3 sheets toasted
 nori seaweed
mayonnaise --- to taste
toasted white sesame seeds
 for coating

1. Make sushi rice according to the instructions on page 195.
2. Cut the nori seaweed in half. Cut the parchment paper into squares that are slightly larger than the nori.
3. Peel and stone the avocado. Cut each one into 12 equal pieces (wedges). Cut the shiso leaves in half.
4. Place the parchment paper onto a *makisu* (rolling mat). Place the nori on top and spread a layer of sushi rice on it. Turn the nori over, so that the nori side faces upwards.
5. A little below the halfway point, make a thin line of avocado strips. Place the shiso leaves on top and spread 1 tsp of mayonnaise thinly over them. Then place the crab sticks on top.
6. Roll the *makisu* up and over the ingredients, pressing it gently. Continue rolling to the edge, making sure you don't roll up the paper in the sushi. Press lightly again, sealed side down, and remove the *makisu* and the paper.
7. Coat the sushi rolls with sesame seeds. Cut each roll into 6 pieces.

牛肉咖喱

Beef curry

　　咖喱是日本极受欢迎的三大家常料理之一。大家经常选用市售的咖喱块来制作咖喱，但在这个食谱里，我会教大家如何用香料熬制一锅香浓的咖喱。洋葱的甜味和浓郁的肉香是美味的关键，所以一定要好好地翻炒洋葱，从而引出食材本身的甜味。

Curried rice is one of the three most popular dishes in Japanese home-cooked meals. It is common to make it with store-bought curry paste, but this recipe takes a more elaborate approach and uses various spices. The sweetness of the onions and the flavor of the meat determine this dish, so make sure to stir-fry the onions thoroughly to pull out their natural sweetness.

牛肉咖喱

[用料4人份]

牛上脑肉（块）1000克
盐 1小匙
胡椒 少许
面粉 3大匙

洋葱 4个（1000克）
胡萝卜 2根（300克）
土豆 4个
番茄 2个（400克）
茴香籽 1小匙
蒜末 1大匙
生姜泥 1大匙
咖喱粉 3大匙

[香料粉]
印度混合香料粉[①]、姜黄粉、香菜粉等
　　各少许
小豆蔻（磨碎）5粒的量

水 6杯（1200毫升）
色拉油 适量
盐、胡椒 各少许
番茄酱 3大匙
伍斯特沙司或炸猪排酱汁 1大匙
米饭 适量
福神渍[②] 适量

1. 洋葱切薄片。
2. 胡萝卜去皮，切圆形或半月形，每块厚度约2厘米。
3. 土豆去皮，切成4等份。泡入清水中以去除部分淀粉，捞出，沥干水分备用。
4. 牛肉切小块，每块边长3～4厘米。
5. 锅中放入2大匙色拉油，小火加热，加入茴香籽，炒香后加洋葱，转中火，翻炒，水分挥发后转小火继续翻炒，洋葱炒至焦糖色后加蒜末和生姜泥，继续炒制。加咖喱粉和混合后的香料粉，粉感消失、产生香味后加水搅匀。
6. 在牛肉块上撒1小匙盐和少许胡椒，揉搓入味，裹面粉。
7. 换一个平底锅，在锅中放入2大匙色拉油，油热后轻煎牛肉块。牛肉块表面均匀地变成焦黄色后，将其放进步骤5的锅中，煮沸后撇去浮沫，盖上锅盖煮炖30～40分钟。
8. 牛肉块变软后，向锅中放入切成大块的番茄。番茄变软后，放入胡萝卜、土豆，继续煮制10～15分钟。所有食材均熟透后，加盐、胡椒、番茄酱、伍斯特沙司或炸猪排酱汁调味。
9. 米饭装盘，浇上咖喱，并根据个人喜好在旁边摆放福神渍。

---
①印度混合香料粉：以丁香、小豆蔻、肉桂为主原料的印度混合香料。
②福神渍：日式月咖喱菜。山萝卜、茄子、莲藕等七种原料腌制成的香菜，因原料品种中由于福神而得名。

Beef curry

[Serves 4]

1kg beef shoulder loin (block)
1 tsp salt
pepper
3 tbsp flour

4 onions (1kg)
2 carrots (300g)
4 potatoes
2 tomatoes (400g)
1 tsp cumin seeds
1 tbsp grated garlic
1 tbsp grated ginger
3 tbsp curry powder

[powdered spices]
garam masala, turmeric, coriander etc.
--- a bit of each
5 cardamom seeds (ground)

6 cups(1200ml) water
vegetable oil
salt, pepper
3 tbsp tomato ketchup
1 tbsp Worcester sauce (or *tonkatsu* sauce)
cooked rice
fukujinzuke pickles
--- to serve

1. Thinly slice the onions.
2. Peel the carrots and slice into 2cm-thick rounds or half-moons.
3. Peel the potatoes and cut them into quarters. Soak them into water, and drain well.
4. Cut the beef into 3-4cm cubes.
5. Heat 2 tbsp of vegetable oil in a pan over low heat, and stir-fry the cumin seeds until fragrant. Add the onion and stir-fry over medium high heat to cook off the moisture, then lower the heat and continue to stir-fry until golden brown. Add the garlic and ginger and stir-fry. Add the curry powder and other spices and stir-fry until they mix well and become fragrant. Add the water and dilute.
6. Sprinkle 1 tsp of salt and a small amount of pepper on the beef and rub them in. Then, coat it with flour.
7. Heat 2 tbsp of vegetable oil in a separate frying pan, and sear the beef. When browned, add it into the pan in step 5. When it comes to a boil, skim the surface, put on a lid, and let it simmer for 30-40 minutes.
8. When the beef is tender, add the coarsely cut tomatoes and cook until they are soft. Add the carrots and potatoes and simmer for another 10-15 minutes. When cooked through, season with some salt, pepper, ketchup and Worcester sauce.
9. Put some rice in a serving bowl and pour some beef curry on top. Garnish with *fukujinzuke* pickles to taste.

茄子干咖喱

Curried rice with eggplant

　　茄子干咖喱是我很喜欢制作的一道菜肴。从我家孩子还小的时候开始，它就经常出现在我家的餐桌上。茄子过油炸一下会更加鲜甜味美。配菜加入了自制腌鸡蛋，这么做出来的鸡蛋空口吃都很好吃，建议大家一定尝试一下。鸡蛋放入冰箱冷藏一晚的话会更加入味。

Curried rice with eggplant has been my favorite recipe since my children were little. The natural sweetness of the eggplant comes out when it is fried. The pickled eggs, for the side dish, taste good by themselves, so please try them. The eggs will taste delicious after being marinated overnight in the fridge, as they will soak in the flavors.

茄子干咖喱

[用料4人份]

猪肉、牛肉混合肉馅 400 克
茄子 4～6 个
洋葱 1 个
青椒 2 个
咖喱粉 2 大匙
咖喱块（市售）2 大匙
香料粉（印度混合香料粉、姜黄粉、
　茴香籽粉、香菜粉等）适量
番茄酱 1 大匙
炸猪排酱汁 1 大匙
煎炸油 适量
色拉油 2 大匙
盐、胡椒 各适量
自制腌鸡蛋 适量
白米饭或杂粮饭、福神渍 各适量

1. 洋葱切小片，每片边长约1厘米。
2. 青椒纵向对半切开，去籽，切小片，每片边长约1厘米。
3. 茄子去蒂，切圆形，每块厚度约3厘米。泡入清水中，捞出，沥干水分备用。锅中倒入煎炸油，放入茄子，茄子炸熟后捞出，沥干油分备用。
4. 向深口平底锅中倒入色拉油，油热后放入混合肉馅，炒至肉馅变色后加入洋葱，翻炒均匀。
5. 向锅中加咖喱粉、咖喱块、香料粉、番茄酱、炸猪排酱汁，炒匀后放入青椒，继续翻炒。关火，加盐和胡椒调味，将炸好的茄子放入锅中。
6. 将温热的米饭盛入盘中，浇上咖喱，旁边摆放自制腌鸡蛋和福神渍。

自制腌鸡蛋

[用料4人份]

鸡蛋 6～8 个
酱油 2 大匙
醋 2 大匙
砂糖 1 小匙

1. 锅中放入鸡蛋，加凉水，刚好没过鸡蛋即可。开火，水沸腾后，用中偏小的火候继续加热。从开火开始计时，煮约12分钟即可关火。捞出鸡蛋，泡入冷水，剥壳，擦干水分。
2. 将酱油、醋、砂糖倒入塑料保鲜袋中，调味汁混合均匀后，放入鸡蛋。
3. 尽可能多地排出塑料袋中的空气，封口。放入冰箱冷藏2~3小时后即可食用。

Curried rice with eggplant

[Serves 4]

400g ground beef and pork mixture
4-6 eggplants
1 onion
2 green peppers
2 tbsp curry powder
2 tbsp curry paste (store-bought, chopped)
powdered spices (garam masala, turmeric, cumin, coriander, etc.) --- a bit of each
1 tbsp tomato ketchup
1 tbsp *tonkatsu* sauce
vegetable oil for deep-frying
2 tbsp vegetable oil
salt, pepper
pickled egg
preferred cooked rice (ex. brown rice)
fukujinzuke pickles --- to serve

1. Cut the onions into 1cm squares.
2. Cut the green peppers in half and remove the seeds. Then chop them into 1cm-square pieces.
3. Trim the eggplants and cut into 3cm-thick round slices. Soak them in water. Drain and pat dry. Heat the deep-frying oil in the frying pan, and deep-fry the eggplants until cooked through. Drain well.
4. Heat the vegetable oil in another deep-frying pan. Add the ground meat and stir-fry. When the meat is browned, add the onions and continue to stir-fry.
5. Add the curry powder, curry paste, spices, tomato ketchup, and *tonkatsu* sauce, and mix. Mix in the green peppers and turn off the heat. Season with salt and pepper, and add the fried eggplants.
6. Ladle the warm curry over warm rice and add the pickled egg and *fukujinzuke*.

Pickled eggs

[Serves 4]

6-8 eggs
2 tbsp soy sauce
1 tbsp vinegar
1 tsp sugar

1. Place the eggs in a pot, cover them with cold water, and heat the pot. When it comes to a boil, turn down the heat to medium low. Let it boil for about 12 minutes from cold water. Then, put the eggs in cold water, peel them, and wipe off excess water.
2. Combine the seasonings in a plastic bag, and soak the eggs in the marinade.
3. Drain the air from the plastic bag, seal it, and let the eggs marinate in the refrigerator for 2-3 hours before eating.

ESSAY 6 随笔

漆器
日本手工艺的珍宝
A Japanese handicraft
that should be cherished
urushi lacquerware

漆 器

　　碗、筷、多层食盒、茶碟、酒器。日本的漆器千千万万。漆器既是日本自古以来的传统工艺，也是日本的象征之一。提到漆器，人们往往会觉得更适合在正月等特别的时刻使用，但其实在我的老家就有很多漆器，例如盛味噌汤的碗、便当盒之类的。因为我妈妈是漆器爱好者。

　　我现在也在认真地保管着一只刻有我的名字的漆器汤碗，自我懂事以来便一直在使用它。随着接触到漆器的美妙，我慢慢地爱上了漆器，有时前去拜访制作者的工作场所，有时自己设计……如此一来，我慢慢地就收集了许多漆器。

　　由于漆器非常"纤细"，所以在保养的过程中有许多注意事项，例如需要小心清洗以防划伤、擦干水分后再晾干等。如果保管、护理得当的话，漆器是可以用一辈子的。另外，美观性自不用说，其实漆器也是实用性满分的容器，比如将漆器用作食物器皿的话，盛装的饭菜既不容易变凉也不容易捂热。现如今年轻人已经很少使用漆器了，此类日本传统手工艺逐渐淡出人们的生活，让人觉得有些寂寞。

　　好的道具会为我们的生活带来更多的快乐，也会使我们的生活变得更加丰富多彩。哪怕进度慢一点，我也希望人们能够重新认识到漆器的优点，并且希望漆器能够再次渗透到人们的日常生活中。出于这种原因，我觉得应该好好地向下一代人传承日本优秀的传统技艺。

漆器便当盒。既可用作便当盒，又可上下分开作为餐盘成对使用。

Urushi lacquer lunchbox. It can be used as a lunchbox or, when separated, a pair of serving dishes.

Urushi

In Japan, there are various shikki (lacquerware goods) such as owan (wooden bowls), chopsticks, jubako (multi-tiered boxes), chataku (Japanese teacup saucers), shuki (sake cups), and many others. Making urushi lacquerware is a traditional Japanese craft and one of the symbols of Japan. People usually think that urushi lacquerware is mainly used for special occasions such as the New Year. But, in my parents' house, my mother loved urushi lacquerware and so we used it in our daily life, for example, bowls for miso soup and lunch boxes.

The lacquered soup bowl with my name on it that I have been using since childhood is my treasure. The more I realize the beauty of urushi lacquerware, the more I love it, and I visit urushi lacquerware craftsmen's workshops, or create urushi lacquerware which I design by myself. I now have quite a collection of lacquerware.

Urushi lacquerware is very delicate so it needs to be appropriately cared for, washed gently, and thoroughly wiped dry. But, if you use it carefully, it will last a lifetime. Besides being beautiful, it is functional in keeping warm food warm and cold food cold. These days, however, it is less used by the younger generation. I feel sad that the works of traditional Japanese craftsmanship, including the urushi lacquerware, are starting to disappear from our daily life.

Beautiful and functional tools can make our life happier and richer. I hope, even if only a little, people realize the greatness of urushi lacquerware and again use it in their daily life. We should never lose the skill required to make this distinguished Japanese handicraft and make sure it is handed down to the next generation.

日式甜蛋卷

Sweet dashi rolled omelet

  这道甜蛋卷看似简单，但实际操作下来会发现，想要把它做好其实是有难度的，煎鸡蛋的火候也需要特别注意。虽说制作甜蛋卷有专用的长方形小煎锅，但只要不在意成品形状，那么用圆形煎锅也是完全没问题的。即使中间的蛋皮煎得不太好看也不要着急，因为后面能够一点一点地调整外形，只要最后一张蛋皮煎得好看就万事大吉。

This dish looks simple, but requires technique in adjusting the heat, and is in fact difficult to make. The square frying pan is exclusively for rolled omelets, but this dish can be made using round frying pans as well, if you don't mind the shape. Even if the layers don't line up during the process, there's no need to worry. If you can get the last layer rolled nicely, it will be a success.

日式甜蛋卷

[用料 1根甜蛋卷]

鸡蛋 6个

[甜口调味汁]

砂糖 40克
日式高汤 1/2杯（100毫升）
淡口酱油 1小匙
盐 少许

色拉油 适量

1. 制作甜口调味汁。向温热的日式高汤中加入砂糖，搅匀使其充分溶解。依次加淡口酱油和盐调味，放凉备用。
2. 在碗中打散鸡蛋，加入甜口调味汁，搅匀后使用滤筛等过滤蛋液。
3. 向蛋卷煎锅中倒入色拉油，加热，用厨房纸巾将色拉油均匀地涂抹在锅中。
4. 倒入一层蛋液，摊平煎制，待表面半熟后，由远离锅柄的一侧向锅柄一侧卷制蛋卷，制成蛋卷内芯。如果油量不足，按步骤3的方法在空出来的地方涂一层薄薄的色拉油即可。再次倒入蛋液，确保蛋液与蛋卷内芯底部接合，卷制蛋卷。多次重复此步骤，直至用尽全部蛋液、蛋卷成形。
5. 自然放凉后，切块，大小方便食用即可。

Sweet dashi rolled omelet

[Makes 1 roll]

6 eggs

[sweet dashi sauce]
40g sugar
1/2 cup(100ml) dashi
1 tsp light soy sauce
salt

vegetable oil

1. Make the sweet dashi sauce: Add sugar in a warm dashi, and dissolve well. Add the light soy sauce, season with some salt, and let cool.
2. Beat the eggs in a bowl. Add the sweet dashi sauce, mix well, and strain.
3. Heat the omelet frying pan, and grease it with vegetable oil using a paper towel.
4. Pour a little egg mixture into the pan. Quickly pull it towards the edge of the pan while it is still half-cooked to create a center roll. If you need to add more oil to the pan, repeat the process in step 3. Add a little more egg mixture into the pan making sure that it flows under the center roll as well. Roll it towards the edge of the pan. Repeat this several times.
5. When it is done, let it cool, and cut it into bite-sized pieces.

什锦拌面

Noodles with shrimp and vegetables

这可是我们家的人气食谱,而且制作十分简单。建议事先做好自制蒜姜酱油备用,这样做炒饭或者拌面的时候可以直接使用,非常方便。另外,只需加上一些香菜和花椒,就可以把味道变成超级正宗的中国风,美味程度令人震惊。

This is one of my family's favorite easy-to-make dishes. The garlic and ginger soy sauce would be nice to have on hand as it can be used for cooking a variety of dishes including fried rice. The added coriander leaves and Szechuan peppercorns give the dish an authentic Chinese taste, and make it surprisingly delicious.

什锦拌面

[用料 4人份]

中式蒸面条 2团
生虾 150克
豆芽 1袋
葱 2根
蒜姜酱油 2大匙
蚝油 1/2大匙
色拉油 适量
绍兴酒或其他酒 1/2大匙
盐、胡椒 各适量
芝麻油 少许
香菜 适量
花椒 适量
辣味醋（醋中加辣椒圈）适量

[蒜姜酱油]

酱油 2杯（400毫升）
大蒜 2~3瓣
生姜 1块
大蒜、生姜切薄片，放入广口瓶，倒入酱油，静置2~3小时。

1. 牛虾剥壳，去虾线。洗净后擦干水分。
2. 豆芽去须根。
3. 葱白切段，每段长约5厘米。纵向将葱段切成薄片。
4. 将缠绕在一起的面团一根一根地、小心仔细地解开。
5. 将蒜姜酱油和蚝油倒入碗中，混合均匀。
6. 平底锅中倒入1/2大匙色拉油，加热，炒制虾，加盐、胡椒、绍兴酒。虾炒熟后盛出备用。
7. 向同一平底锅中倒入1大匙色拉油，倒入面条，炒熟，盛出备用。
8. 再向同一平底锅中倒入1~2大匙色拉油，大火炒豆芽和葱。
9. 将虾和面条倒回锅中，快速搅匀。关火，淋上步骤5制作的调味汁，加盐、胡椒、芝麻油调味。
10. 装盘，撒适量花椒粉（可自行研磨），添加香菜和辣味醋。

Noodles with shrimp and vegetables

[Serves 4]

2 portions steamed Chinese noodles

150g shelled shrimp

1 pack bean sprouts

2 Japanese leeks

2 tbsp garlic and ginger soy sauce

1/2 tbsp oyster sauce

vegetable oil

1/2 tbsp *shokoshu* (Chinese sake) or Japanese sake

salt and pepper

sesame oil

coriander leaves for garnish

Szechuan peppercorns, chopped red chili, vinegar --- to serve

[garlic and ginger soy sauce]

2 cups(400ml) soy sauce

2-3 cloves garlic

1 knob ginger

Slice the garlic and ginger thinly. Put the garlic and ginger in a jar and add the soy sauce. Let it stand for over 2 to 3 hours.

1. Rinse the shrimp and devein them if any. Pat them dry.
2. Remove the tails from the bean sprouts.
3. Cut the leeks into 5cm-long pieces and then slice thinly.
4. Gently pull the noodles apart one by one to loosen.
5. Combine the garlic and ginger soy sauce and oyster sauce in a bowl.
6. Heat 1/2 tbsp of oil in a frying pan and stir-fry the shrimp. Add the salt, pepper and *shokoshu*. Remove them from the pan once they are cooked.
7. Add 1 tbsp oil to the pan, add the noodles, and stir-fry. Remove them from the pan.
8. Add 1-2 tbsp oil to the pan and stir-fry the bean sprouts and leeks on high heat.
9. Put the shrimp and noodles back into the pan and stir. Turn off the heat and add the previously made sauce and toss them together. Add salt, pepper, and sesame oil to taste.
10. Pile onto a serving plate and sprinkle with ground Szechuan pepper and garnish with coriander leaves. Serve with vinegar mixed with chopped chili.

笼屉荞麦面

"Zaru soba"
(cold soba noodles)

日式蘸面汁

Mentsuyu (noodle broth)

用开水煮熟、冷水冲洗的荞麦面鲜嫩爽滑，自家制作的日式蘸面汁清新爽口。再加上葱、绿芥末、海苔碎这几样简单的食材调味。这就是我非常喜爱的日式味道。这个蘸面汁的做法是我从母亲那里学来的。我们不仅可以用它给荞麦面、乌冬面调味，也可以用它来烹饪炖菜、盖饭，所以真的非常建议大家学一下这个做法。

Soba, boiled and washed under cold running water, served with *mentsuyu*. The only garnishes are leek, wasabi and crumbled nori seaweed. I like the simple style of this dish. My mother taught me the recipe of this *mentsuyu*. It can be used not just for noodles, but also for many recipes such as simmered dishes or *donburi* (a bowl of rice topped with meat or vegetables and sauce). So this recipe will surely come in handy.

笼屉荞麦面

[用料 2人份]

荞麦面（干）200克
日式蘸面汁 适量
海苔碎 适量
绿芥末泥 适量
葱 适量
（切成葱花，过一遍凉水，捞出）

1. 锅中倒入大量清水，烧开后，放入荞麦面。偶尔搅拌以防面条粘连。
2. 水开后加入1杯凉水，继续煮至面条熟透。
3. 用漏勺捞出面条，用流水充分冲洗后，沥干水分。
4. 装盘，撒海苔碎。用其他容器盛装日式蘸面汁。添加绿芥末泥和葱花。

日式蘸面汁

[用料 易于制作的分量]

凉水 4杯（800毫升）
酱油 1+1/2杯（300毫升）
甜料酒 1杯（200毫升）
砂糖 40克
鲣鱼花 50克

1. 小锅中倒入4杯凉水，加酱油、甜料酒、砂糖，搅拌均匀使调料充分溶于水。
2. 开火，在水快要沸腾的时候放入鲣鱼花，转中火继续加热2～3分钟，水再次沸腾后关火。自然放凉。
3. 过筛后放入冰箱冷藏。
   *可冷藏保存4～5天。

"Zaru soba" (cold soba noodles)

[Serves 2]

200g dried
　soba noodles
mentsuyu
crumbled nori seaweed
grated wasabi
Japanese leek
　(chopped and soaked
　in water)
　--- for garnish

1. Bring plenty of water to a boil and add the soba. Stir the noodles occasionally to prevent them from sticking.
2. When the noodles are about to boil over, add 1 cup of water and continue boiling until the noodles are cooked through.
3. Drain and wash the noodles well under cold running water. Drain them well.
4. Put the noodles in serving plates and sprinkle crumbled nori seaweed on top. Pour the *mentsuyu* in separate serving bowls. Garnish with wasabi and Japanese leek.

Mentsuyu (noodle broth)

[Ingredients]

4 cups(800ml) water
1+1/2 cups(300ml)
　soy sauce
1 cup(200ml) mirin
40g sugar
50g bonito flakes

1. Combine the water, soy sauce, and mirin in a pan. Add the sugar and dissolve it.
2. Heat it until just before boiling. Add the bonito flakes and simmer over low heat for a few minutes. Remove from the heat and let it stand until it slightly cools.
3. Strain through a sieve and chill in the refrigerator.
   * *Mentsuyu* keeps in the refrigerator for 4-5 days.

煎 饺

Gyoza
(Chinese dumplings)

　　煎饺是我特别喜爱的料理之一。煎饺馅料的搭配多种多样，其中我最喜欢的便是接下来要讲的这个配方。注意在煎制饺子的时候要倒入适量的面粉水，这样能把底部煎得更加松脆。趁热吃堪称一绝。建议大家在蘸料中多放点姜丝，再加入适量的酱油、醋、辣油，会更好吃。

Gyoza is one of my favorite dishes. There are various fillings for *gyoza*, but this is my favorite recipe. By adding the water with a little flour when braising the *gyoza*, the bottoms get crispy. It's delicious to eat while they are still piping hot. Enjoy them with plenty of ginger, soy sauce with vinegar and chili oil.

煎 饺

[用料 24 个]

猪肉馅 150 克

白菜 150 克

卷心菜 150 克

韭菜（切碎）50 克

蒜末 1 大匙

绍兴酒 1 大匙

汤（用 1 大匙热开水冲开 1 小匙中式汤粉后放凉）

芝麻油 适量

色拉油 少许

盐、胡椒 各适量

饺子皮 24 张

面粉 1 小匙

清水 1/2 杯（100 毫升）

酱油、醋、辣油 各适量

姜丝 适量

1. 白菜、卷心菜切细丝。分别放入两个碗中，各加 1 小匙盐，拌匀后静置一会儿，渗出水分后捞出，挤干水分备用。
2. 平底锅中放入 1/2 大匙芝麻油，油热后放入蒜末，适当翻炒，注意不要炒煳。
3. 碗中依次放入猪肉馅、绍兴酒、汤，混合均匀。将步骤 2 炒香的蒜末和芝麻油一起倒入碗中，加白菜、卷心菜、韭菜，搅拌均匀后，加盐和胡椒调味。用保鲜膜包上碗口，腌制约 30 分钟。
4. 包饺子。用勺子等工具挖取适量馅料（步骤 3）置于饺子皮上，在饺子皮边缘涂抹少许清水，捏褶，封口。
5. 将 1 小匙面粉放入碗中，倒入 1/2 杯清水，搅拌均匀。
6. 分两锅煎制饺子，每次煎一半。将平底锅置于炉上，热锅后加少许色拉油，呈圆形摆放饺子，煎一会儿。倒入一半面粉水（步骤 5），盖上锅盖，用偏小的中火焖煎一会儿。水分基本挥发后打开锅盖，淋少许芝麻油，煎至饺子底部酥脆即可。装盘，将煎饺倒扣在盘子上，煎至焦黄色的底部朝上。用同样的方法煎制另一半饺子。将酱油、醋、辣油、姜丝混合在一起，搭配蘸汁食用煎饺。

Gyoza (Chinese dumplings)

[Makes 24]

150g ground pork
150g *hakusai* (Chinese cabbage)
150g cabbage
50g finely chopped *nira* (Chinese chives)
1 tbsp chopped garlic
1 tbsp *shokoshu* (Chinese sake)
soup (1 tsp Chinese soup paste dissolved in 1 tbsp hot water and cooled)
sesame oil
vegetable oil
salt and pepper
24 thin *gyoza* wrappers (dumpling wrappers)
1 tsp flour
1/2 cup(100ml) water
soy sauce, vinegar, chili oil ginger (cut into fine trips) --- to serve

1. Finely chop the *hakusai* and cabbage. Put them into separate bowls. Sprinkle 1tsp of salt into each bowl. Mix both of them lightly, and let stand for a while. Squeeze water out of the vegetables.
2. Put 1/2 tbsp sesame oil in a frying pan, and fry the chopped garlic.
3. Put the ground pork in a bowl. Add the *shokoshu* and soup, and mix together. Follow with the fried garlic, along with the oil, add the *hakusai*, cabbage, and *nira* and mix well. Season with some salt and pepper. Cover with plastic wrap and let it stand for 30 minutes.
4. Scoop the mixture onto a *gyoza* wrapper with a knife or spoon. Put the mixture on each *gyoza* wrapper. Wet the edges with a little water, fold it over, and pinch to seal.
5. Dissolve the flour with the water.
6. Pan-fry half of the dumplings at a time. Heat a frying pan and add some oil. Arrange the dumplings in a circle and cook for a short time. Pour half of the water with flour, cover with a lid, and cook over low-medium heat. When the water has almost evaporated, pour a little sesame oil around the dumplings, and cook until the bottoms of the dumplings get crisp. Turn over onto a plate and serve with soy sauce, vinegar, chili oil, and ginger.

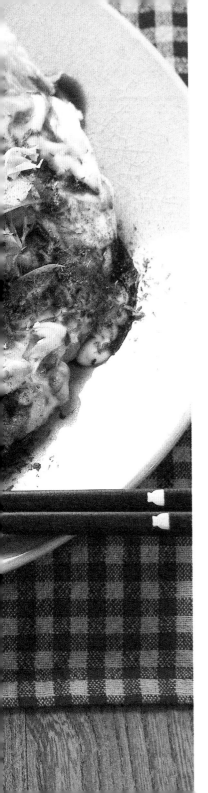

日式什锦煎饼

Okonomiyaki
(savory Japanese pancakes)

　　日式什锦煎饼是极受外国友人欢迎的日式菜肴之一。除了本食谱中介绍的墨鱼、鲜虾、猪肉外，您还可以尝试放入自己喜欢的任何食材。建议先用微波炉（200瓦）将鲣鱼花加热5～6分钟后，再将其撒在什锦煎饼上，这样处理的鲣鱼花口感松脆，味道更佳。另外，对什锦煎饼来说，日式红姜也是必不可少的配菜。

Okonomiyaki (literally 'your favorite things grilled') is one of the Japanese dishes that foreign people love. You can use any ingredients you like in place of the squid, shrimp, or pork that I introduce here. If you microwave (at 200W) the dried bonito flakes for about 5 to 6 minutes, it will become crispy and more flavorful. Sprinkle it over the *okonomiyaki* and add a little red pickled ginger for a delicious taste treat.

日式什锦煎饼

[用料 4人份]

墨鱼（躯干部分）1只（120克）
鲜虾 15只（150克）
猪肋条肉片（火锅用）200克
卷心菜叶 3片（150克）
日式红姜 2～3大匙
小葱葱花 1/2杯
（炸天妇罗时形成的）面衣碎渣 2～3大匙
鸡蛋 4个
色拉油 适量

[面糊]

山药 100克
鸡蛋 1个
日式高汤 1杯（200毫升）
面粉 1杯（200毫升）

[顶部配料]

什锦煎饼酱汁、蛋黄酱、青海苔碎、鲣鱼花 各适量

1. 墨鱼切块，每块边长约2厘米。
2. 鲜虾去壳去尾，放平，水平方向对半切开，取出虾线，再切至长度的一半。
3. 猪肉切段，每段边长4～5厘米。
4. 卷心菜叶切细丝，日式红姜切细丝。
5. 山药去皮，擦成泥。
6. 将蛋液在碗中打散后，加入日式高汤和山药泥，混合均匀。加入面粉，轻轻搅匀。加入卷心菜、小葱葱花、面衣碎渣、红姜，搅匀，最后加入墨鱼和鲜虾。
7. 平底锅中倒入少量色拉油，油热后倒入1/4的面糊（1人份），并将猪肉片展开平铺在上面，煎制3～4分钟。翻面，继续煎3～4分钟至食材熟透。
8. 另起一个平底锅，倒入少量色拉油，油热后打入一个鸡蛋。戳破并轻轻搅动蛋黄部分。将步骤7制作的煎饼放在鸡蛋上，注意煎饼贴有猪肉片的一面朝下，与煎蛋接合。
9. 装盘，煎蛋面朝上。依次挤上什锦煎饼酱汁和蛋黄酱，最后撒上青海苔碎和鲣鱼花。按同样步骤制作剩余3张什锦煎饼。

Okonomiyaki (savory Japanese pancakes)

[Serves 4]

1 squid (body part, 120g)
15 shrimp (150g)
200g sliced pork belly
  (for *shabushabu*)
3 cabbage leaves (150g)
2-3 tbsp red pickled ginger
1/2 cup chopped spring onion
2-3 tbsp *agedama*
  (leftover fried tempura bits)
4 eggs
vegetable oil

[pancake batter]
100g *yamaimo* yam
1 egg
1 cup (200ml) dashi
1 cup (200ml) flour

[topping]
mayonnaise
okonomiyaki sauce
  (a thick Worcester sauce)
aonori (green laver flakes)
dried bonito flakes

1. Cut the squid into 2cm-square pieces.
2. Remove the shells and tails from the shrimp. Slice them horizontally in half. Devein them, and cut in half lengthwise.
3. Cut the pork slices into 4-5cm strips.
4. Shred the cabbage. Chop the red ginger finely.
5. Peel and grate the *yamaimo*.
6. Mix 1 egg, dashi, and grated *yamaimo* in a bowl. Add the flour and stir lightly. Add the shredded cabbage, chopped spring onion, *agedama*, and red ginger, and combine. Add the squid and shrimp.
7. Heat a little oil in a frying pan. Pour in 1/4 of the batter (for 1 serving). Put the pork slices on top and cook for 3-4 minutes. Flip the pancake over and cook till ready (another 3-4 minutes).
8. Heat a little oil in another frying pan. Crack an egg into the pan. Stir the yolk lightly. Place the pancake from step 7 on top of the egg, pork side down, and continue cooking.
9. Turn it over onto a serving plate and spread *okonomiyaki* sauce and mayonnaise on top. Sprinkle with *aonori* and bonito flakes. Make 3 more *okonomiyaki* with the remaining batter in the same way.

金枪鱼面包小点

Tuna crostini

腌金枪鱼是从古代传承至现代的传统日式菜肴,既可以直接食用,也可以搭配面包食用。因此,我尝试着开发了一道新菜品,搭配牛油果,把腌金枪鱼这道传统和食融入经典意大利菜头盘面包小点①中。酱汁里的蒜香是美味的关键。

Soy sauce-marinated tuna has been enjoyed in Japan since long ago. It's tasty as it is, but it also goes well with bread, so I arranged it with avocado into a sort of an Italian appetizer. The garlic added in the sauce makes it especially savory.

---

①面包小点:英文写作"crostini",经典意大利菜头盘,一般由一小片烤面包和芝士、肉类、蔬菜等配料组成。

金枪鱼面包小点

[用料4人份]

金枪鱼（寿司用，红肉或中鱼腩肉）1块（150克）
大蒜1小瓣
酱油2大匙
酒2大匙

[牛油果酱]
牛油果1个
柠檬汁 少许
橄榄油1大匙
盐、胡椒 各少许

黄瓜 1/2根
法棍切片（每片厚度2厘米）8～10片

1. 大蒜切薄片。
2. 锅中加水，加热。水烧开后放入金枪鱼，煮15秒。捞出，立即泡入冰水。冷却后用厨房纸巾擦干水分，动作尽量谨慎。
3. 拿出一个储存容器，向其中倒入酱油和酒，混合均匀后分别放入金枪鱼和大蒜，放进冰箱冷藏腌制2～3小时。食用前擦干腌汁，切块，每块厚度5～8毫米。
4. 牛油果对半切开，去核，用勺子挖出果肉放入碗中，压成泥。加柠檬汁和橄榄油，拌匀，用盐和胡椒调味。
5. 纵向将黄瓜削成薄片。
6. 加热烤网，烤制法棍切片至适当变色。将牛油果泥涂满面包，放上黄瓜、金枪鱼（擦去腌汁）、蒜片。可根据个人喜好再刷一层橄榄油。

Tuna crostini

[Serves 4]

1 block (150g) tuna
(for sashimi / lean meat or medium fatty flesh)
1 small clove garlic
2 tbsp soy sauce
2 tbsp sake

[avocado cream]
1 avocado
squeezed lemon juice
1 tbsp olive oil
salt, pepper

1/2 cucumber
8-10 slices baguette
(each 2cm-thick)

1. Cut the garlic into thin slices.
2. Bring a pot of water to a boil, put the tuna in and boil for 15 seconds. Take out the tuna and immediately soak in ice water to cool. When cooled, drain, and wipe thoroughly with a paper towel.
3. Combine the soy sauce with sake in a food storage container, marinate the tuna in it, top it with garlic slices and chill in the refrigerator for 2-3 hours. Drain the tuna and cut into 5-8mm-thick slices just before eating.
4. Halve and stone the avocado. Using a spoon, scoop out the flesh into a bowl and mash. Add lemon juice and olive oil and mix. Season with salt and pepper.
5. Thinly slice the cucumber lengthwise.
6. Heat the grill and toast the baguette until crispy brown. Apply avocado cream evenly to the surface, and lay the cucumber slices. Top it with drained tuna and garlic slices. Sprinkle olive oil if preferred.

免揉面包
No-knead bread

我对刚从烤箱中拿出来的热乎乎的现烤面包情有独钟。哪怕自己烤制的面包并不完美,但是能趁热吃到热乎乎的面包就会让人觉得很幸福。制作这个面包比较花时间,但是因为不需要揉面团,所以制作起来比较简单。大家也可以根据个人喜好,尝试在面团中加上芝士、葡萄干、坚果等配料。

I love piping hot bread, fresh out of the oven. This is the advantage of baking your own bread. Even if it's not done perfectly, you can still enjoy the bliss of eating a freshly baked one. This recipe is a bit time-consuming, but there's no need to knead, so it doesn't require special skills. Feel free to experiment with the recipe by adding different ingredients to the dough, such as cheese, raisins, and nuts.

免揉面包

[用料 1个直径25厘米的球形面包]

高筋面粉 300克
砂糖 2小匙
盐 1/2小匙
速溶干酵母 1小匙
水 200毫升 （水温约25摄氏度）
干粉 （高筋面粉）适量

1. 碗中倒入高筋面粉，分别加入砂糖、盐、速溶干酵母后搅拌均匀。加水，用木勺搅拌至食材无粉感且成团后用保鲜膜包住碗口，静置醒面约2小时（第一次发酵）。面团膨胀至原始大小的2倍左右为宜。
   *气温较低（冬季等）时面团所需发酵时间较长，因此也可以通过烤箱的发酵功能，设置成30～35摄氏度发酵即可。

2. 面团膨胀后在碗壁和面团之间撒上适量干粉，用卡片从容器边缘取出面团并用双手拿住。不要将面团放在台面上。用手掌将面团整形至一团，用双手将面团延展开，再团成团，再用双手延展开，注意每次拉伸开面团的时候都需要向面团中添加适量干粉，多次重复此过程以排出面团中的空气。待面团表面光滑、呈圆球形、底部无开口裂缝后，即可将其放入碗中，封上保鲜膜，静置发酵约1小时（第二次发酵）。

3. 锅中铺上烘焙纸，再次按步骤2的方法整形面团，在面团表面轻轻地撒上干粉后盖上盖子，再次静置发酵约1小时，直至锅中面团膨胀至1.5倍为止（第三次发酵）。

4. 烤箱预热至200摄氏度。

5. 把锅放在烤盘上，不要打开锅盖，烤制约30分钟。面团充分膨胀后打开锅盖，继续以200摄氏度烤30分钟即可。面包表皮变色后，取出面包，置于烤网上自然放凉。另外，可以根据个人口味，趁热在刚出锅的面包上涂抹黄油等调味品。

No-knead bread

[1 boule 25cm diameter pot]

300g bread flour
2 tsp sugar
1/2 tsp salt
1 tsp instant dry yeast
200ml water (around 25°C)
bread flour for dusting

1. Put the bread flour into a bowl, add sugar, salt, dry yeast separately in that order, and mix well. Pour in the room-temperature water, and stir well with a wooden spoon until all the flour mixture is incorporated. Form into a ball in the bowl, cover with plastic wrap and let rise for about 2 hours. (primary fermentation) The dough should rise to about double the original size.
   * When the room temperature is low, such as during the winter season, fermentation takes longer. You can use the fermentation function of the oven at around 30-35°C
2. When the dough has risen, dust some flour between the dough and the bowl, scrape out the dough with dough scraper from the edges of the bowl, hold the dough in both hands. Shape the dough into a ball in your palms, stretch it out, and shape it into a ball again, while dusting more flour. Repeat this process several times to let the gas out. Smooth out the surface and shape into a ball. Return the dough to the bowl seam-side down, cover with plastic wrap and let stand for about an hour. (secondary fermentation)
3. Line the pot with parchment paper. Repeat the process in step 2 and shape the dough into a ball. Dust the surface with flour and place it into the pot and put on the lid. Let it rest in the pot for about another hour until the dough rises to one and a half times the original size. (third fermentation)
4. Preheat the oven to 200°C.
5. Place the pot in the oven and bake for 30 minutes with the lid on. When the dough has risen sufficiently, remove the lid and bake for another 30 minutes at 200°C. When the surface turns golden brown, take it out of the oven onto a rack and let it cool down. Serve with butter etc. fresh out of the oven.

猪肉蔬菜味噌汤

Pork and vegetable miso soup

猪肉蔬菜味噌汤日语写作"豚汁"，是日本几乎无人不知、无人不爱的传统家常菜肴。做这道料理的时候能够用到各种各样的切菜方法，所以我在教小朋友做饭的时候也经常会选择这道料理。

This is a favorite dish of the Japanese. You can learn various ways to cut vegetables, so I often choose this dish when teaching cooking to children.

猪肉蔬菜味噌汤

[用料4人份]

白萝卜5厘米（200克）
胡萝卜5厘米（100克）
牛蒡1/2根（100克）
蒟蒻1块（200克）
芋头2～3个（180克）
猪肋条肉薄切片 200克
日式高汤6杯（1200毫升）
味噌 5～6大匙
葱 少许
（切成葱花，过一遍凉水，捞出）
七味辣椒粉 适量

1. 白萝卜、胡萝卜去皮，切成银杏叶形，每片厚度约5毫米。
2. 牛蒡去皮，削成薄片后，将其泡入凉水以去除涩味。捞出，沥干水分。
3. 把蒟蒻弄成一口大小，焯水以去除碱味，用漏勺捞出，沥干水分后备用。
4. 芋头去皮切成圆形或半月形后，将其泡入凉水，捞出，沥干水分备用。
5. 猪肉切至一口大小。
6. 向锅中倒入日式高汤，加热，放入白萝卜、胡萝卜、牛蒡、蒟蒻、芋头。煮沸后撇出浮沫，转小火。
7. 蔬菜变软后将猪肉片下锅。猪肉片煮熟后放入味噌。
8. 装盘，撒上葱花。可根据个人喜好，加入七味辣椒粉调味。

Pork and vegetable miso soup

[Serves 4]

5cm daikon(200g)
5cm carrot(100g)
1/2 burdock(100g)
1 piece konnyaku(200g)
2-3 taroes(180g)
200g thinly sliced pork ribs
6 cups(1200ml) dashi
5-6 tbsp miso
Japanese leek, chopped and soaked in water, *shichimi* pepper --- for garnish

1. Peel the daikon and carrot, cut them in quarters lengthwise, and slice them into 5mm-thick quarter-rounds(*icho-giri*). (see page 280)
2. Peel the burdock and shave it into *sasagaki* slices. (see page 284) Soak the burdock in cold water to remove the bitterness, then drain.
3. Tear the konnyaku into bite-sized pieces. Boil it in water to remove the bitterness, then drain.
4. Peel the taro and cut horizontally into *wagiri* or *hangetsu-giri* pieces (rounds or half-moons). Soak in water, then drain.
5. Cut the pork into bite-sized pieces.
6. Heat the dashi in a pot, and add the daikon, carrot, burdock, konnyaku and taro. Skim the surface when the broth comes to a boil. Lower the heat.
7. When the vegetables become soft, add the pork. When the pork is cooked through, dissolve the miso into the broth.
8. Pour in a serving bowl and garnish with chopped leek. Serve with *shichimi* pepper to taste.

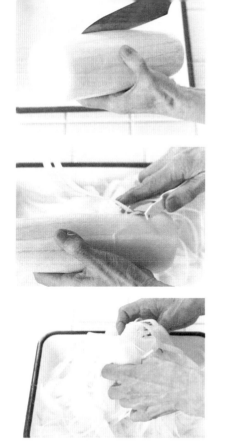

用削皮器将萝卜削成面状细条,分量要足。放入一次刚好能吃完的量,煮好之后一定要趁热食用。可根据个人口味调整加热时间,既可以简单加热保留萝卜的爽脆口感,又可以多煮一会儿将萝卜炖烂入味,不管怎么吃都很美味。这道菜品口味比较清淡,蔬菜吃再多也不会腻。

Shave the daikon with a peeler into plenty of noodle-like strips. Put only the amount you can eat at one time into the pot, and enjoy it while it's still piping hot. You can enjoy the texture of the daikon if you take it out of the soup quickly. It also tastes good when it has been simmered well. It is lightly flavored, so you can eat plenty of vegetables without feeling heavy in the stomach.

暖身萝卜锅
Daikon noodle pot

暖身萝卜锅

[用料4人份]

猪肉（火锅用猪肉卷）
（上脑/肋条等部位）
300～400克
萝卜 长约20厘米（净重约500克）
水芹菜 3把

[汤汁]
日式高汤8杯（1600毫升）
淡口酱油 2大匙
酱油 2大匙
酒 2大匙
盐 1小匙

柚子 适量
七味辣椒粉 适量
山椒粉 适量

1. 萝卜去皮，在每纵向间隔1.5厘米的地方划一刀后，用削皮器将萝卜削成面条状细长条。
2. 洗净水芹菜，沥干水分。去掉底部较硬的部分，切成两半。
3. 锅中倒入汤汁所需材料，开火，煮沸后将萝卜下锅，萝卜煮熟后放猪肉卷，注意将猪肉卷展开，一片一片地依次放入锅中，防止肉片卷边或粘连。最后加入水芹菜，简单加热一下，烫熟即可。关火后连汤汁一起倒入碗中，可根据个人喜好添加七味辣椒粉或山椒粉调味。

Daikon noodle pot

[Serves 4]

300-400g pork
 for *shabushabu*
 (shoulder loin/rib etc.)
20cm piece daikon
 (about 500g)
3 bundles watercress

[broth]
8 cups (1600ml) dashi
2 tbsp light soy sauce
2 tbsp soy sauce
2 tbsp sake
1 tsp salt

yuzu
shichimi pepper
sansho powder --- to taste

1. Peel the daikon, make incisions along the length of the daikon at 1.5cm intervals, then shave the daikon using a peeler to make noodle-like strips.
2. Wash the watercress and drain well. Trim the woody stem ends, and cut in half.
3. Combine the ingredients for the broth in a pan and heat it. When it comes to a boil, add the daikon and let it cook, and then add the pork while spreading out each piece. Finally, add the watercress just before turning off the heat. Put it in a serving bowl, squeeze some yuzu over it, and garnish with *shichimi* or *sansho* powder if preferred.

莲藕、牛蒡之类的根茎蔬菜，小松菜、水菜之类的青菜，芹菜、阳荷之类的香辛蔬菜，柚子、酸橘之类的柑橘类植物……我喜欢的食材实在是太多了，要真一个一个列举起来就没完没了了。刚刚列举的食材都是我非常喜爱的日式味道，也是我做饭时不可或缺的材料。其中，柚子是我每年最期待的应季味道。到了柚子的季节，我便将柚子全方位地融入我的料理之中。例如，我会像挤柠檬一样，直接将柚子汁挤进各种荤素菜品里。柚子皮也不能浪费。虽说只需将一点点柚子皮放入料理之中就可以起到非常明显的提味效果，但是每到柚子的季节，我都会将柚子皮切成末，然后把它们大量地加到菜品中。要我说，这正是每年柚子季才能享受到的奢侈体验。另外，柚子果汁其实也另有妙用。柚子榨汁后和酱油混合制成的柚子酱油非常美味且用途广泛，可以用于调制火锅酱料，也可以给拌菜和炒菜提味。柚子酱油是我们家四季常备、不可或缺的调料之一，所以每年到了柚子季，我都会做好一整年分量的柚子酱油，把它们放进冷藏柜和冷冻柜分别保存。

　　随着日式料理的人气日益高涨，现如今在世界各地都可以买到各式各样的日式食材，但即便如此，要在国外买到这些应季食材还是有一定难度的。每次主要面向海外群体介绍日料食谱的时候，为了能够让大家都可以亲自下厨尝试，我都会尽量选用一些容易买到的常见食材。可是，另一方面，我还是希望能够让大家更加了解日式食材的优点和美味。如果大家有机会入手这些食材，希望大家能够多多尝试，感受下它们的魅力。

[柚子]

柑橘类的一种。分为成熟后采摘的黄柚（秋冬上市）以及成熟前采摘的青柚（初夏至初秋上市），以其独特的清爽香气为特征。柚子果肉极酸，日本人基本不会直接生食，而是选择将其果汁制成调味料，增加酸味和清香。在日料中，将切薄的柚子果皮放入菜品中提味也是很常见的用法。果皮内侧的白色部分稍带苦味，因此巧用柚子皮时，直接用刀从果实外侧削下一层薄薄的果皮即可。另外，冬天泡澡的时候也可以在浴池里加入一些柚子皮，泡一个香气四溢的柚子浴。

Root vegetables including lotus root and burdock, green vegetables including *komatsuna* and *mizuna*, flavorful vegetables including *seri* and *myoga*, and citrus fruits including yuzu and sudachi – I have so many favorite Japanese ingredients. They are all essential to my recipes.

Among them, yuzu is the flavor whose season I look forward to the most. During its season, I directly squeeze yuzu on meat, fish, and vegetables as you would a lemon. Even a tiny zest of yuzu can enhance the original taste of the dish, but I enjoy the luxury of using plenty of shredded yuzu zest when it's in season. Yuzu soy sauce, a mixture of yuzu juice and soy sauce, is an important seasoning in my family – we use it for all sorts of dishes, from dipping sauce for hot-pots to vegetable dressing and stir-fried food. During its season, I make enough yuzu soy sauce for one year and preserve it in the refrigerator and freezer.

The more Japanese food has become popular internationally, the easier it has become to find Japanese ingredients overseas. However, such seasonal ingredients as yuzu remain difficult to obtain. When I introduce recipes to people living outside Japan, I usually arrange the recipes with ingredients they can easily obtain, because I hope as many people as possible will try and enjoy this style of cooking without any difficulty. But on the other hand, I hope more and more people realize how delicious these Japanese seasonal ingredients are. I would be happy if you should find an opportunity to taste them and experience the pleasure of these seasonal foods.

....

【 Yuzu 】

Yuzu is a kind of citrus fruit. There are two types of yuzu: yellow yuzu, from autumn to winter, is harvested when the fruit is ripe; and green yuzu, from early summer to early autumn, is harvested before the fruit is ripe. Yuzu has a unique fresh aroma. The flesh is so sour that it is not eaten, but the juice is used as a seasoning to add aroma and tartness to dishes. Thinly shaved slices of the zest are also used for flavor. As the white part beneath the zest is very bitter, only the surface of the zest is shaved and used. This fruit is not only for cooking – people also put yuzu in their bathtub and enjoy its aroma in wintertime.

4

DESSERTS

甜品

我的工作是从每天的茶点时间开始的。自从事电视台幕后工作的时候开始,我就一直想着如果有朝一日成为一名料理家,一定要好好享受每天早上的茶点时间。在为了不知何时到来的那一天而努力的日子里,我反复练习的就是这种戚风蛋糕。现在制作这道甜品的时候我也会忆起当年,一边庆幸自己坚持努力到了现在,一边决心将它做得更加美味。

My work day begins by enjoying a cup of tea and some sweets with my coworkers. I started out as a staff member working behind the scenes for a TV cooking show. But even back then, I had been telling myself that I would cherish morning teatime if I ever became a cooking writer. It was this chiffon cake that I had been practicing many times in anticipation of that day. Even now, when I make this cake, it reminds me of those days, and I feel glad that I have come this far. At the same time, it makes me more determined to make it even better.

调味戚风蛋糕
Chiffon cake

调味戚风蛋糕

［用料1个直径24厘米的大号蛋糕］

低筋面粉 2杯（约220克）
发酵粉 1大匙
大号鸡蛋 10个
细砂糖 1杯（约180克）
色拉油 1/2杯（100毫升）
水 1/2杯（100毫升）

[调味料]

肉桂粉 1/2～1大匙
多香果 1小匙
丁香 1小匙
葛缕子籽 2大匙

糖粉 适量

[准备工作] 烤箱预热至170摄氏度。

1. 分离蛋清和蛋黄。
2. 在蛋黄中加入一半细砂糖，使用打蛋器打发至黏稠。
3. 先后加入色拉油和水，仔细搅拌。
4. 将低筋面粉和烘焙粉混合后过筛，用打蛋器轻拌，随后加入各种香料继续搅拌，直至面糊无粉粒、变顺滑。
5. 制作蛋白霜。将蛋白倒入另一个碗，使用手持电动搅拌器打发至六分发，加入剩余细砂糖，继续打发至偏硬状态。
6. 向步骤4制作的面糊中倒入1/3的蛋白霜（步骤5），使用橡皮刮刀充分搅拌，分两次倒入剩余的蛋白霜，快速搅拌至颜色均匀。
7. 从较高处将面糊倒入模具。
8. 在台面上轻震两三次模具，以消除面糊中的多余空气，减少气泡。
9. 放入烤箱，以170摄氏度烤制45～55分钟。出炉后倒扣模具，并用容器支撑模具中央，以防模具内空气不流通。
10. 冷却后脱模。插入抹刀，分别沿模具侧面和底部切割，剥离蛋糕，小心不要损伤模具。
11. 切块装盘，可根据个人喜好撒上一些糖粉。

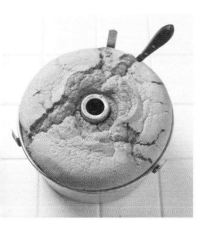

Chiffon cake

[24cm tube pan : makes 1]

2 cups (220g) flour
1 tbsp baking powder
10 eggs (large)
1 cup (180g) granulated sugar
1/2 cup (100ml) vegetable oil
1/2 cup (100ml) water

[spices]
1/2-1 tbsp cinnamon powder
1 tsp allspice
1 tsp clove
2 tbsp caraway seed

powdered sugar

[preparation] Preheat the oven to 170°C.

1. Separate the eggs into yolks and whites.
2. Add half of the granulated sugar into egg yolks and beat well with a whisk until it thickens.
3. Add first the vegetable oil and then the water and mix well.
4. Sift together the flour and baking powder, mix lightly with a whisk. Add the spices and mix well until all the flour mixture is incorporated and the batter is smooth.
5. Make the meringue: In a separate bowl, beat the egg whites with a hand mixer until soft peaks form. Add the rest of the granulated sugar and beat until it forms hard peaks.
6. Add a third of the meringue in step 5 into the egg yolk mixture in step 4. Fold with a rubber spatula. Repeat the same process with the rest of the meringue, and fold quickly until the color of the mixture is uniform.
7. Pour the batter into the tube pan from a slightly higher place.
8. Lift the pan and drop onto the counter several times to help release air bubbles from the batter.
9. Bake for 45-55 minutes in a preheated 170°C oven. Take it out of the oven and flip the pan over to let the cake cool completely upside down. Support the center hole of the pan with a container so as not to let the cake sweat.
10. Carefully run a knife around the edges making sure not to scratch the pan, and take the cake out of the pan gently.
11. Cut the cake into serving-size pieces, and serve on a plate. Sprinkle with powdered sugar to taste.

鲜奶冻
Panna cotta

鲜奶冻是我家的传统甜品，一做就是好多年。稍带点朗姆酒风味，是大人喜欢的味道。吉利丁用量点到为止，能够使鲜奶冻凝固即可，这样可以确保鲜奶冻入口即化，口感绵软。可以说没人不爱这种味道。

I have been making this dessert for a long time, and it is one of my family's favorites. The rum gives it a more adult flavor. I use minimum amount of gelatin to give it a creamy texture that melts in your mouth. This is a dessert everyone loves.

鲜奶冻

[用料 小号8~10个]

牛奶 1＋1/2 杯（300毫升）
香草豆荚 1/2 根
细砂糖 60 克
吉利丁粉 1 袋（5 克）
水 2 大匙
生奶油 1 杯（200毫升）
朗姆酒 2 大匙

[焦糖浆]

细砂糖 50 克
水 1 小匙
热水 1/4 杯（50毫升）

1. 在小碗中加水后，加入吉利丁粉，待其凝为胶状。
2. 将牛奶倒入锅中。纵切香草豆荚并取出豆子，将豆荚和豆子一同放入锅中以小火加热。加入细砂糖，充分溶解。
3. 水即将沸腾时关火。取出香草豆荚，将胶状的吉利丁全部倒入锅中，搅拌使其溶解。
4. 少量多次加入生奶油，搅拌均匀，加入朗姆酒调味。
5. 将锅底置于冰水中，冷却锅中液体的同时，将其搅拌至稍微黏稠。
6. 倒入容器，放入冰箱冷藏。
7. 制作焦糖浆。将细砂糖和水倒入小锅，小火加热。熬至糖浆稍显颜色后轻晃小锅，直至变为焦糖色关火。关火后加热水，待自然冷却后，将其倒入鲜奶冻。

Panna cotta

[Makes 8 to 10 small ones]

1+1/2 cups (300ml) milk
1/2 vanilla bean
60g granulated sugar
1 pack (5g) gelatin powder
2 tbsp water
1 cup (200ml) heavy cream
2 tbsp rum

[caramel sauce]
50g granulated sugar
1 tsp water
1/4 cup (50ml) hot water

1. Sprinkle gelatin powder evenly over 2 tbsp water in a small bowl, and let it soften.
2. Put the milk in a pan. Split the vanilla bean pod lengthways and scrape out the seeds. Put the pod and the seeds into the pan and cook over low heat. Add granulated sugar and let it dissolve.
3. Turn the heat off just before it comes to a boil. Take out the vanilla pod, and add all of the softened gelatin. Mix and let it dissolve.
4. Add the heavy cream little by little while mixing. Add flavor with a spritz of rum.
5. Prepare a bowl full of ice water, cool the pan on top of it, mix and cool until it slightly thickens.
6. Pour the mixture into each container and chill in the refrigerator until it sets.
7. Make the caramel sauce: Put the granulated sugar and water in a small pan and heat. When the color starts to change, shake the pan and cook over low heat until it turns light amber. Turn off the heat, and add the hot water. When it cools, pour the sauce over the panna cotta.

小仓冰激凌
Ogura ice cream

　在蛋香十足的香草冰激凌中加入市售熟小豆便制成这种日式和风甜品。只要记住香草冰激凌的做法，就可以通过更换配料来制作各种口味的冰激凌，用草莓、蓝莓、木莓等水果替代小豆即可。

This is a Japanese-style dessert, made by adding store-bought boiled azuki beans to vanilla ice cream that contains lots of egg. All you have to do is to learn how to make the basic vanilla ice cream. Then, you will be able to enjoy various flavors by adding fruit, such as strawberries, blueberries, and raspberries instead of azuki beans.

小仓冰激凌

[用料 易于制作的分量]

鸡蛋 2个
细砂糖 50克
生奶油 1杯（200毫升）
粒状红豆馅（市售）100克
熟小豆（市售）200克

1. 将鸡蛋打入碗中搅拌均匀，加入一半的细砂糖后使用打蛋器仔细搅匀。
2. 将生奶油和剩余的细砂糖放入其他碗中，打发至偏硬状态。
3. 将步骤1的蛋液倒入步骤2打发的奶油中，搅拌均匀后，倒入平底盘或其他容器中，包好保鲜膜，放入冰箱冷冻。在其快要凝固的时候，加入粒状红豆馅和熟小豆。随后重新将其放进冰箱冷冻室，在彻底冷冻前再搅拌2～3次即可。

Ogura ice cream

[Ingredients]

2 eggs
50g granulated sugar
1 cup (200ml) heavy cream
100g sweet azuki bean paste with skin (store-bought)
200g boiled azuki beans (store-bought)

1. Beat the eggs in a bowl, add half of the granulated sugar and mix well using a whisk.
2. Pour the heavy cream into a separate bowl, add the rest of the granulated sugar and whisk until you reach stiff peaks.
3. Add the mixture in step 1 into step 2. Pour the mixture into a shallow container, cover with plastic wrap and put in the freezer. As it starts to freeze, take it out of the freezer and mix in the azuki bean paste and the beans, put it back in the freezer. Stir the mixture several times while it freezes.

妈妈甜甜圈

Mom's doughnuts

这是以前妈妈教给我的一道甜点,充满了我们的回忆。母亲很少制作甜点,但是因为父亲喜欢,以前她便经常给我们做这个。这个方子做出来的甜甜圈非常松软,十分美味。

My mother taught me how to make these doughnuts a long time ago. And it brings back memories. My mother didn't make many sweets, but she often made these doughnuts because they were my father's favorite. The very loose dough is the secret to making surprisingly delicious doughnuts.

妈妈甜甜圈

[用料 环型、封闭型甜甜圈各10个]

鸡蛋 2个
砂糖（上白糖）80克
牛奶 1/4杯（50毫升）
黄油 40克
低筋面粉 250克
烘焙粉 2小匙
干面粉（高筋面粉）适量
煎炸油 适量
砂糖（上白糖，用于收尾）适量
肉桂粉 适量

1. 融化黄油。将黄油放入小号耐热容器，包好保鲜膜后，放入微波炉（600瓦）加热约30秒。
2. 将鸡蛋打入碗中，使用打蛋器拌匀后，加入砂糖继续搅拌。
3. 加入牛奶和融化了的黄油，继续搅拌。
4. 将低筋面粉、烘焙粉混合后筛入碗中，用橡胶刮刀搅拌均匀至面团顺滑无颗粒。
5. 在烘焙纸上撒适量干面粉后，将步骤4制作的面团置于其上，揉成厚度约1.5厘米的形状后，在其表面撒适量干面粉。封好保鲜膜，放入冰箱冷藏，醒面3小时以上至面团变硬。
6. 取出面团，用模具压成型，每次都需蘸适量面粉防粘。因面团柔软，所以要注意力度，小心操作。
7. 煎炸油加热至约180摄氏度时将甜甜圈下锅，为防止炸煳需多次翻面炸制，每个甜甜圈需恰到好处地油炸2～2.5分钟，使其内外均匀受热。按炸熟顺序，依次取出甜甜圈。
8. 收尾。基本版只需在出锅后立刻撒上足量的砂糖。肉桂版则需在撒匀砂糖后继续撒上适量肉桂粉。

Mom's doughnuts

[Makes 10 ring doughnuts and 10 doughnut holes]

2 eggs
80g sugar
  (refined white sugar)
1/4 cup (50ml) milk
40g butter
250g flour
2 tsp baking powder
bread flour for dusting
oil for deep-frying
sugar (refined white sugar), cinnamon powder
  --- for sprinkling on top

1. Melt the butter by placing in a small heat-resistant container, cover loosely with a plastic wrap, and microwave (600W) for about 30seconds.
2. Crack the eggs in a bowl. Whisk them, add the sugar and mix well.
3. Add the milk and melted butter and continue to mix well.
4. Sift flour and baking powder into the mixture, and mix well with a rubber spatula until smooth.
5. Place the dough on step 4 on floured parchment paper. Roll it out to a thickness of 1.5cm. Dust some more flour from the top, and loosely cover it with a plastic wrap. Let it rest in the refrigerator for more than 3 hours until the dough hardens.
6. Take out the dough from the fridge, cut with a floured doughnut cutter. Flour the cutter each time. The dough is very soft, so treat with care.
7. Heat oil to around 180°C, and deep-fry the doughnuts. Turn them over so as not to overcook them, and deep-fry for 2 to 2-and-a half minutes until they are golden brown and cooked through. Take out the smaller doughnut holes that are done first.
8. Topping: For the plain doughnuts, sprinkle plenty of sugar while still hot. For the cinnamon ones, sprinkle cinnamon powder after they have been sprinkled with sugar.

松软松饼
Fluffy pancakes

这道松饼中加入了足量的蛋白霜，口感十分松软。虽说打发蛋白稍微有点花时间，但是非常值得。我之前在夏威夷的大学里教过这道甜品，当时大家反馈说头一次吃到如此好吃的松饼，我记得自己当时特别开心。

These pancakes are especially fluffy because of the added meringue. Whipping up the egg whites is a bit of work, but it's definitely worth it. When I taught this recipe at the college in Hawaii, everybody told me that they had never tasted pancakes this good. I remember how happy I was to hear that.

松 软 松 饼

[用料 直径12厘米 4张]

原味酸奶（无糖）
　1/2杯（100毫升）
鸡蛋 2个
细砂糖 30克
牛奶 1/4杯（50毫升）
低筋面粉 100克
烘焙粉 1小匙
色拉油或黄油 少量
黄油 适量
喜欢的果酱 适量
枫糖浆 适量

1. 在碗上放置筛网，铺上厨房纸巾，倒入酸奶。包好保鲜膜，放入冰箱冷藏1小时以上，去除多余水分。
2. 分离蛋白和蛋黄。
3. 向蛋黄中加入1/3的细砂糖，用打蛋器充分搅匀。随后加入去除多余水分的酸奶和事先准备好的牛奶，继续搅拌。
4. 向蛋白中加入剩余的细砂糖，用手持电动打蛋器打发至偏硬状态。
5. 低筋面粉和烘焙粉混合过筛，筛入步骤3的混合物中，用橡胶刮刀拌匀。
6. 将步骤4打发的蛋白霜的1/3加入步骤5调制的面糊中，搅拌均匀后，再加入剩余的蛋白霜，快速切拌以防消泡。
7. 向平底锅中倒入色拉油或黄油，油热后，倒入1/4的面糊。调小火力，盖上锅盖，松饼边缘烤好后翻面烤制，保证内外受热均匀。用相同的方法烤制另外3张松饼。
8. 将刚出锅的松饼盛入盘中，按个人喜好搭配黄油、各式果酱、枫糖浆。

Fluffy pancakes

[Makes four 12cm-diameter ones]

1/2 cup (100ml) plain yogurt (unsweetened)
2 eggs
30g granulated sugar
1/4 cup(50ml) milk
100g flour
1 tsp baking powder
vegetable oil or butter
butter, jam, maple syrup --- for toppings

1. Place a colander in a bowl and line them with a paper towel. Put the yogurt in it and cover with plastic wrap. Let it stand for at least an hour in the refrigerator to drain the yogurt.
2. Separate the eggs into yolks and whites.
3. Add a third of the sugar to the yolks and mix well with a whisk. Add the drained yogurt and milk and mix well.
4. Add the remaining sugar to the egg whites and whisk with a hand mixer until it makes stiff peaks.
5. Mix the flour and baking powder and sift it into the mixture in step 3. Fold with a rubber spatula.
6. Add a third of the mixture in step 4 into step 5, and mix. Add the remaining mixture and fold gently so as not to deflate the egg whip.
7. Heat some vegetable oil or butter in a frying pan, pour in a quarter of the batter. Lower the heat and put on a lid. When the edges are cooked, flip it over and cook through. Repeat this process with the remaining batter.
8. Serve fresh out of the pan with toppings of your choice, such as butter, jam or maple syrup.

零失败芝士蛋糕
Fail-proof cheese cake

人们经常觉得制作甜点难度较高，这个芝士蛋糕却简单到只需将所有材料混合在一起就可以进行烤制了。对于甜品初学者来说也十分友好。我经常用院子里的花草制成的小花束来装饰这个蛋糕，然后把它送给亲朋好友。

People tend to think that baking sweets is difficult. But with this cheese cake, all you have to do is to mix the ingredients into one bowl and bake. It will come out delicious, even if you are baking it for the first time. I often give it out as a present, decorating it with a tiny bouquet of herbs from my garden.

零失败芝士蛋糕

[用料 直径18厘米的圆形蛋糕1个]

奶油干酪1盒（200克）

黄油30克

（有盐、无盐均可）

饼干

（全麦类）100克

细砂糖1/2杯（90克）

鸡蛋2个

生奶油1杯（200毫升）

低筋面粉3大匙

柠檬汁1大匙

糖粉 适量

[准备工作]

将奶油干酪放入碗中，室温软化。

同时，室温软化黄油。

在模具的底部和侧面包好烘焙纸。

烤箱预热至160～170摄氏度。

1. 将饼干放入食品袋，用研磨棒等压碎后，倒入黄油，混合均匀。安装好模具的活动底，外翻食品袋将饼干碎倒入模具，铺满底部，按压紧实。
2. 将奶油干酪放入碗中，使用手持电动搅拌器混合至顺滑。
3. 依次加入细砂糖和鸡蛋，搅拌均匀。
4. 加入生奶油，搅拌至浓稠。
5. 筛入低筋面粉，使用橡胶刮刀轻拌，加入柠檬汁后继续搅拌。
6. 将面糊倒入模具，手持模具外围，在台面上轻震两三次，以消除多余空气。放入预热好的烤箱中烤制40～45分钟。
7. 取出晾凉后脱模，放入冰箱冷藏。
   *烤制好的芝士蛋糕可冷冻保存。
8. 切块装盘，可根据个人喜好撒上一些糖粉。

Fail-proof cheese cake

[18cm round cake pan with removable bottom / makes 1]

1 box (200g) cream cheese

30g butter (salted or unsalted)

100g biscuits (whole-wheat)

1/2 cup (90g) granulated sugar

2 eggs

1 cup (200ml) heavy cream

3 tbsp flour

1 tbsp lemon juice

powdered sugar --- for topping

[preparation]

Place the cream cheese in a bowl and bring to room temperature.

Bring the butter to room temperature.

Line the bottom and sides of the pan with parchment paper.

Preheat the oven to 160-170°C.

1. Put the biscuits into a plastic bag and coarsely crush them with a pestle (or rolling pin), and mix in the butter. Empty the bag in the cake pan. Flip the plastic bag inside out, put your hand inside and press the crumbs evenly and firmly to the bottom of the pan.
2. Beat the cream cheese in the bowl with a hand mixer until soft and smooth.
3. Add granulated sugar and eggs in that order and mix well.
4. Add the heavy cream and mix well till it thickens.
5. Sift in the flour and fold lightly with a rubber spatula. Add the lemon juice and mix some more.
6. Pour the batter into the pan. Holding the edge of the pan, gently lift and drop the pan several times on the counter to let the air out. Bake for 40-45 minutes in a preheated oven.
7. Remove from the oven and set aside to cool. Once it has cooled, remove from the pan and cool in the refrigerator.
    * Baked cheese cake can be kept in the freezer.
8. Slice it into serving-size pieces, and sprinkle powdered sugar to taste.

蔬菜的切法

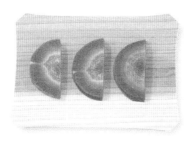

半月形切法
Hangetsu-giri
(Half-moons)

银杏叶形切法
Icho-giri
(Quarter-rounds)

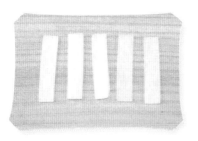

长方形切法
Tanzaku-giri
(Rectangles)

How to cut vegetables

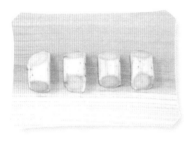

滚刀块切法
Ran-giri
(Random-shaped)

梳形块切法
Kushigata-giri
(Wedges)

半月形切法

将蔬菜切成半圆形,因半圆形似半月而得名。适用于胡萝卜、白萝卜、芜菁等圆柱形的蔬菜。先将食材竖着对半切开,再从两端开始切。切至适当厚度即可。

Hangetsu-giri (Half-moons)

Cutting ingredients into half-moons. Used for cutting cylindrical vegetables such as carrot, daikon radish and turnip. Cut vegetables into 2 lengthwise pieces and slice them crosswise to achieve the desired thickness.

银杏叶形切法

将蔬菜切成扇形,因扇形形似银杏叶而得名。适用于胡萝卜、白萝卜、芜菁等圆柱形的蔬菜。先竖着将食材切成4等份,再从两端开始切薄片。

Icho-giri (Quarter-rounds)

Cutting ingredients into quarter-rounds in the shape of a ginkgo leaf. Used for cutting cylindrical vegetables such as carrot, daikon radish, and turnip. Cut the vegetables into 4 lengthwise pieces and slice them crosswise.

长方形切法

将蔬菜切成薄薄的长方形,形似长条诗笺。先将食材切至想要的长度和宽度的长方体,再从两端开始切薄片。炒菜和炖汤的时候经常需要长方形的蔬菜,也经常用这种方法来切割海苔。

Tanzaku-giri (Rectangles)

Cutting ingredients into thin rectangles. Cut ingredients into pieces of the desired length and width, then slice them into thinner pieces. Suitable for cutting nori or vegetables for stir-fried dishes and soups.

滚刀块切法

将食材切成不规则形状。滚动食材,切断纤维。用滚刀块切法切成的食材的大小取决于制作的菜肴种类,但注意同一菜肴中食材的大小应尽量保持一致。滚刀块切法能够使食材的表面积增大,更容易熟,也更容易入味。适合做炖菜、煲汤。

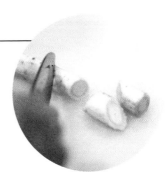

Ran-giri (Random-shaped)

Cutting ingredients into random-shaped but even-sized pieces. Cut the ingredient diagonally against the grain and rotate it. This ensures a large surface area of the ingredients, so you can cook these random-shaped wedges rapidly and season them thoroughly. Suitable for cutting ingredients for simmered dishes and soups.

梳形块切法

将洋葱、番茄、柠檬等球形食材呈放射状切开,因切块形似梳子而得名。先将食材竖着对半切开,再从中心向外侧均等切块。

Kushigata-giri (Wedges)

Cutting spherical ingredients such as onion, tomato, and lemon into wedges. Cut the ingredient into 2 lengthwise pieces and cut them radially into even-sized pieces.

蔬菜的切法

削 切

Sogi-giri
(Shaving cut)

片

Sasagaki
(*Sasagaki* shavings)

How to cut vegetables

切葱末

Fine chopping
(for Japanese leeks)

葱白切丝

Shiraga-negi
(White hair leek)

削切

将食材削成薄片。将菜刀放平,从外侧向自己的方向切,即从远离刀柄的一侧向刀柄一侧切。适用于需要食材厚度一致或增大食材表面积的情况。

Sogi-giri (Shaving cut)

Shaving ingredients thinly. Place a knife flatly against the ingredient and slice it by pulling the knife toward you. Suitable to ensure the even thickness of pieces or to cut pieces with a large surface area.

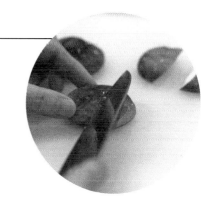

片

一边旋转食材,一边像削铅笔一样从一端开始削薄片。适用于牛蒡等棒状食材。

Sasagaki (*Sasagaki* shavings)

Sasagaki is one way to slice vegetables and is used for shaving long and narrow vegetables like Japanese burdock root. Keep the vegetable in your hand and while you rotate it, shave or slice thinly with a knife as if sharpening a pencil.

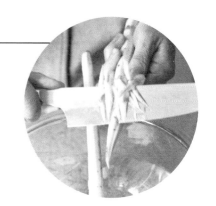

切葱末

大葱切末,需要先用刀尖在大葱表皮留下密集的划痕,再从一端开始切碎。如果想要切得更碎一点,可以用刀拍。

Fine chopping (for Japanese leeks)

Make scores in a Japanese leek with the tip of a knife. Slice it crosswise into small pieces. If necessary, chop these into smaller pieces.

葱白切丝

将葱白切成细丝。先将葱白切成4厘米长的葱段,再将葱段纵向切开,取出葱芯。展开外侧白色的部分,铺平,沿着纤维纹路将其切成细丝。将切好的葱丝在清水中浸泡一会儿,捞出,沥干水分备用。葱丝可用作面食或其他菜肴的配菜。

Shiraga-negi (White hair leek)

Cutting the white part of a Japanese leek into julienne strips. Cut a Japanese leek into 4cm-long pieces. Score the pieces lengthwise, open them, and remove the core. Unfold and pile the remaining layers of the leek and cut along the grain into julienne strips. Soak them in water, drain, and use as a garnish for noodles and other dishes.

职员表

摄影：耕田头、莫妮卡·魏斯曼、迪雷克·牧岛
造型：福泉响子
设计：粟辻设计（粟辻美早、仁科悦子）
合作执笔：浅野未华
英文翻译：小岛绘里子
合作摄影：田岛刚（株式会社vex）
校对：中泽悦子
编辑：汤原一宪（NHK出版）

图书在版编目（CIP）数据

我开动了！栗原晴美的美味料理笔记／（日）栗原晴美著；牛丹迪译．—武汉：华中科技大学出版社，2020.6
ISBN 978-7-5680-6074-5

Ⅰ.①我… Ⅱ.①栗… ②牛… Ⅲ.①菜谱-日本 Ⅳ.①TS972.183.13

中国版本图书馆CIP数据核字（2020）第058828号

Harumi
© 2018 Harumi Kurihara
Originally published in Japan in 2018 by NHK Publishing, Inc.
Chinese (Simplified Character Only) translation rights arranged with NHK Publishing, Inc.
through TOHAN CORPORATION,TOKYO.

本作品简体中文版由NHK Publishing授权华中科技大学出版社有限责任公司在中华人民共和国境内（但不含香港、澳门和台湾地区）出版、发行。

湖北省版权局著作权合同登记　图字：17-2020-035号

我开动了！栗原晴美的美味料理笔记
Wo Kaidong le Li Yuan Qing Mei de Meiwei Liaoli Biji

［日］栗原晴美 著
牛丹迪 译

出版发行：华中科技大学出版社（中国·武汉）	电话：(027) 81321913
北京有书至美文化传媒有限公司	(010) 67326910-6023

出 版 人：阮海洪

责任编辑：莽　昱　刘　韬
责任监印：徐　露　郑红红　封面设计：邱　宏

制　　作：北京博逸文化传播有限公司
印　　刷：北京华联印刷有限公司
开　　本：889mm×1194mm　1/32
印　　张：9
字　　数：36千字
版　　次：2020年6月第1版第1次印刷
定　　价：89.00元

本书若有印装质量问题，请向出版社营销中心调换
全国免费服务热线：400 6679 118　竭诚为您服务
版权所有　侵权必究